CONCRETE FIT FOR PEOPLE

A Practical Introduction to a Bio-Functional Eco-Architecture for the Third Millennium A.D.

U.K.	Pergamon Press Ltd., Headington Hill Hall, Oxford OX3 0BW, England
U.S.A.	Pergamon Press Inc., Maxwell House, Fairview Park, Elmsford, New York 10523, U.S.A.
CANADA	Pergamon of Canada, Suite 104, 150 Consumers Road, Willowdale, Ontario M2J 1P9, Canada
AUSTRALIA	Pergamon Press (Aust.) Pty. Ltd., P.O. Box 544, Potts Point, N.S.W. 2011, Australia
FRANCE	Pergamon Press SARL, 24 rue des Ecoles, 75240 Paris, Cedex 05, France
FEDERAL REPUBLIC OF GERMANY	Pergamon Press GmbH, 6242 Kronberg-Taunus, Hammerweg 6, Federal Republic of Germany

First edition 1980

British Library Cataloguing in Publication Data

Ritter, Paul
Concrete fit for people.
1. Concrete 2. Polystyrene
3. Plastics in building
I. Title
721'.0445 NA3695 79-40711

ISBN 0-08-024671-0

Printed in Great Britain by A. Wheaton & Co. Ltd., Exeter

A Practical Introduction to a Bio-Functional Eco-Architecture for the Third Millennium A.D.

CONCRETE FIT FOR PEOPLE

PAUL RITTER

Ph.D(Soc.f), Ph.D(Soc.f), MCD, B.Arch., FRIBA, MRTPI, FRAIA, MRATPI, MACE

Director P.E.E.R. Institute,
Councillor of the City of Perth, capital W. Australia,
Formerly
City Architect and Planner of Perth;
and
Member of the Council of the Royal Institute of Architects, London.

Published by the Down to Earth Bookshop Press, Perth, Western Australia.
This edition distributed by
Pergamon Press, Oxford, New York, Toronto, Sydney, Paris, and Frankfurt.

Contents

Dedicated to my wife and partner

For photographs in this book I want to thank
Henry Bartnik, Jean Ritter, Ernest Polis, Ralph Hibble, Leonora Ritter, Paul Ritter.
For drawings Witold Generowicz.

Acknowledgements are an integral part of this book. If I have forgotten any one the loss is mine as well as theirs and I humbly apologise.

I INTRODUCTION

Concrete has covered much of the urban earth. The drab, dull and deadening, streaky and blotchy blight is hated universally. It is one of the cruelties humanity has inflicted upon itself. it is not the fault of the concrete. Transformation is possible. We can add delight to the materialistically utilitarian attributes of building to give all those qualities with which the name architecture should be associated. Concrete can be, should be, fit for people.

This book is the tip of an iceberg. It is the entire bit of visible evidence of a large loose body of technology, floating about in deep waters, whipped up to choppy confusion by currents of architectural whims and fashions. To take the simile a little further, we may add that the accurate assessment of what the immense ice mountain will be like, when it emerges, is obscured by the water's optical distortions.

But it will not melt away. The tip is just one sign of the coming climate.

Humanity is growing sensitive to itself as never before. The emerging era, the third millenium, will have as its most basic, destinguishing feature not the conquering of outer space, not the development of the incredibly effective, electronic gadgetry, but the new sensitivity, awareness and insights into our emotions. At last there beacons to us an effective diagnosis and understanding of the human illness which has, so far, by and large, brought to nought, the noble human intentions. It has blocked the national use of our creativity and resources and has tabooed the richest and liveliest aspects of life. 'The best things in life have been free' — but unavailable.

In a life-positive, inspiring school of architecture in Los Angeles I listened to top experts on space cities. They made vivid the technical brilliance that made us, right now, capable of hurling the building materials into place from the weaker gravity fields of the moon. I was astounded. But, with others I wondered, and asked, "why space cities?" The answer was that it was a natural development of man's expansion, due to current difficulties on earch, just as people had emigrated to the USA when things became difficult in Europe. This made me think that if only humanity had been able to sort itself out prior to leaving Europe for the USA then the desecration of so much of that part of the earth may have been avoided. The space city, as described and despite its technological wizardry, perpetuates the insensitivities still normal in our emotionally ill societies, west and east. I have a hope that we may heal ourselves sufficiently, in time, not to infect other worlds or space cities.

Concrete Fit For People springs from this desire and the newly found possibilities for physicians and architects "to heal themselves", in the emerging era, the era of sensitivity, the third millenium. The technical superabundance of the current computor-drunk years will become a life-positive force. The era will harness, define and effectively channel natural over-reactions of those striving for ill-defined freedoms, rights and privileges, withheld for far too long, from women, workers, adolescents, children, babies, all of us, withheld by ourselves from ourselves.

This book deals with techniques that may hasten a renaissance of architecture by including the sociological and educational aspects that can make it happen. The pattern of ideas and actions in **Concrete Fit for People** has grown from the same roots as **Planning for Man and Motor,** (1000 illustrations) i.e. "planning fit for people"; **Educreation** (500 references) "education fit for people"; **The Free Family** (400 pages) "bringing up kids as if they were people". In each case the second edition adds to text and title ". . . and Feedback". Users report back.

Sensitive, creative observations and assessment are the rare contributions towards improvements of ideas and inventions. To have experienced difficulties while actually using and working with concepts gives the right to criticise. Most communication, including criticism, in our emotionally ill society, springs alas, from the need to let out, aggression, characteristic of our sick society, and not from a desire to help or improve. Watch for it, restrain it, I try to, constantly.

II BIO-FUNCTIONAL ECO-ARCHITECTURE

II.1. Our era.

The bio-functional eco-architecture will spring from the awareness of our era. The past few thousand years have brought architecture based on the dire need to keep out the forces of nature, destructive to humans, and to provide luxury and beauty, for the few as we excelled at this.

As I see it the next millenium, (only ten 100 year old people laid end to end) will progressively, and with the help of bio-technic skills and manufacture, produce not only the protection against nature, but a new harmony with nature. As the species has spawned across the earth we have clustered more and more in cities. They used to be culturally and ecologically balanced, or insignificant, agglomorations of individuals until some 300 years ago. However now their nature is different. Their effect on ecology, on human health and culture is devastating and progressively so. The bio-functional counter trend in harmony with nature is already under way: e.g. the cleansing of London's river Thames and industrial skies of Sheffield, U.K. Pittsburgh, U.S.A. etc. The indications are clear. We will be 'friends of the earth'. An architecture which takes this human intention seriously is inevitable. Burying the buildings, one approach suggested, seems pretty negative and drastic. But little by little the intentions will be translated into bio-technology and style. Something more than just "solar houses". We suggest some of the fundamental sources for the development of the bio-functional style, before we illustrate our contemporary, crude, early attempts.

This era is of evolutionary significance for our species. The settlement on this earth will be guided by the **quality** we desire, and the need to be healthy and creative. Compulsive **quantitative** growth, as a means for survival, as is usual in nature, is being superceded. We will ensure survival through qualitative rather than quantitative criteria.

II.2. Philosophical background.

The contents of this book make most sense in the context of my philosophy. An inkling of this can be gained by looking at the meaning of the few words I have coined, inspite of my antipathy for doing that sort of thing. But these are needed to obviate clumsy lengthy, descriptive phrases.

Common Functioning Principles.
The term first coined by Wilhelm Reich in the fifties, indicates a way that shows the unity of the world, the way it all "hangs together". Too many have ignored the simple fact that it must hang together or it would fall apart. As simple as that. A basic energy, orgone, ("or by any other name"), permeates the world as we know it. It functions according to certain principles. They are **common** functioning principles, as the basic energy pervades all. To look for them, as we try to understand things, gives us valuable insights which are complementary to and extend knowledge that we already have.

This hypothesis has proved so useful in Reich's work and in mine, as well as that of others, that it can be recommended. An interesting corollary has been noted by us: wherever fundamental research takes an honest look, even in complete ignorance of the Common Functioning Principles concept, they are noted and recorded, described. That they **are** Common Functioning Principles is usually ignored. Science does not currently encourage connections between fields, but on the contrary specialisation.

Attraction — Fusion — Liberation describes the three-beat, creative rhythm, the most basic of the principles.

The tendency towards **spiral motion,** in the broadest sense, is another, as is the tendency to **branching.**

As with the chemical elements the discovery of the whole set may take some time.

Opportunion is another — the tendency to seek ways and means to carry out functions — rather than stay, grumbling and blaming, stuck and frustrated. The new word is needed as the word 'opportunism' has been degenerated into meaning 'taking advantage' of another, unfairly, rather than of actions which benefit all.

Emotil — this is my simple name for the emotional illness, specific to our species. The simple name indicates the simple nature of its origin as my explanation shows: Homo sapien is sensitive. When emotions are stressed and strained beyond an individuals, or a groups, elastic limit a survival mechanism comes into operation. As the lizard looses his tail to stay alive, so homo sapien cuts off feelings with muscle and mental tensions that remain as "blocks" or "armour", in mind or body, defending him from feeling too much ever again.

Alas this throws his emotional and energy functioning into imbalance. The primary, natural balance is replaced with a secondary system. It is this which explains why the species is cruel, unfeeling, sadistic, masochistic, suffers from inertia and is generally unable to gain the ideals it seeks and manages to spell out quite clearly. We can now recognise this illness for what is it: common, ordinary and devastating in its universal occurance. This is basic to ridding the race of this plague. Understanding raises levels of tolerance. As the very young are most vulnerable to emotil the perpetuation of it through the accepted modes of bringing up children, both east and west, is easily understood. Awareness of the nature of emotil, attempts at prevention and healing, springing from this, realisation are widely in use (and misuse). Remember, everything that happens in this world will be affected by emotil.

Emoplay — the word, coined by my daughter Leonora, denotes the everyday activity of recognising that emotil will occur everywhere, and modifying ones actions accordingly. Unreasonable, obstructionist, aggressive reactions, if forseen, can be minimised in their effect. To know that they are most likely to occur when new ideas, initiative, creativity, vitality, reason, sincerity or enthusiasm disturb a status quo, is being forarmed because forwarned. (see my book THE TEN TABOOS, in preparation). For the young to have this realisation as part of their essential education would help against, nihilism, scepticism of the negative sort, despair and the blind hatred of what is a pretty difficult world to understand and cope with without some such insight as spelled out above.

The truth are statements about the healthy; **the countertruth** is factual and true because it describes the ill world, suffering from emotil, the world as we all know it. If you learn to emoplay well, then you can sort out the truth from the countertruth and seeking opportunities, you function better. An example of the truth is that "it is possible to build to satisfy the emotionally sensitive homo sapiens of this earth, one and all.

The countertruth, which applies to the world now, is that "it is not possible to build in such a way because the world is administratively, politically etc etc. incapable of doing it.

Educreation. 'Education for creation growth and change'. An approach which unifies learning and teaching, thinking and doing, with a therapeutic attitude, (emoplay) fitting to a world that is emerging from emotil symptions.

Teamology, the study of doing things more effectively together then apart. Taking seriously the fact that we are a **social** species with special sensitivity, potential and problems. For example we are the only species that faces bringing up 'litters' of different ages, at the same time. Each individual makes sense only as seen as part of relationships.

Eco-architecture, architecture designed to merge creatively with an ecological pattern.

Participation sculture, the creation of sculpture with public, artistic participation, for public buildings, or sculpture designed to allow participation of the public by inviting play, aggression outlets, exploration, etc.

II.3. The Origins of a new style of Architecture.

What do we want from a building material for the twenty-first century, the next millenium? The biotechnic era? The generations of the last of the kings Louis in France, and during the reigns of Victoria and Edward in Britain, were concerned with columns rather than kitchens, formal composition and symmetry rather than comfort and sincerity. Buildings had culminated into an over-rich hotchpotch expressing the materialistic excesses of the upper classes and their desire to demonstrate their status. In the twentieth century reactions and new ideas came from many sources.

Le Corbusier, the French architect who probably had a greater influence on the profession than any other in the early twentieth century, decreed 'a house is a machine for living in'. Like the Bauhaus and others he was concerned to change the then current, non-functional approach.

The first functional phase has now taken over worldwide. But as the twentieth century matured so did the view of the pattern of functions to be satisfied. The bio-energetic approach of our bio-technic era no longer sees the human being as a machine, nor the house as 'a machine for living in'. The ecological relationship between man and his surroundings is understood better. We have become aware and more sensitive towards every aspect of our environment. The house is an extension of the people who live in it, in an organic and subtle sense. We are beginning to learn the hundred and one considerations that apply with such an approach and how much we could achieve with highly developed technology created by the talents of ours and the previous era, and the endless permutations and combinations of these.

This realization slides smoothly out of the basic work on bio-energy by Wilhelm Reich M.D. (1927-57), L.L. Whyte's **The Next Development in Man** (1942), Alfred North Whitehead **Nature and Life,** (1912), Burley Griffin's little-known articles (1921) developing Frank Lloyd Wright's ideas (1910), Pettigrew's enormous work **'Design in Nature, illustrated by Spiral and other arrangements in the Inorganic and Organic Kingdoms as exemplified in Matter, Force, Life, Growth, Rhythms, and specially in Crystals, Plants and Animals'** (1902). He was the last of the great 'natural scientists' who saw the world as one, and also one of the first to give this oneness a meaning that creates a superb bridge from the specialization era to the new era of unifying concepts, now gaining ground all round.

In the second half of the twentieth century we have seen the basic content and theories of these and many related approaches verified and extended over and over again.

In **Educreation** (1966)* I have described some of this more fully, and quoted the undisputed scientific greats as saying, one and all, that we need a significantly fresh departure to gain fundamental springboards for advances in each of their fields. And so it has been in architecture.

The crude analysis of what we might require from building materials in the bio-energetic era is offered as a beginning. We may have advanced two squares in our impatience, and urgency towards an improved 'space ship' earth, (as Buckminster Fuller named it: A master piece of awareness). The species found, contrary to its beliefs for 7000 years of its existence, that the moon and the stars were quite easy to reach, but not a cure for our emotional illness or sick passions.

With 'feeling' as part of our scientific, analytical, and inventive methods, our creative and constructional activities, more bio-energetic needs of organisms will be detected and require sensitive satisfaction. That is a basic distinguishing feature of the bio-energetic era.

Just how much of verified experience and work awaits to be unified and made respectable can be gleaned from the 1500 references in Professor Tromp's book 'Psychical Physics', and many, many other sources. The origin of the new architecture lies in the feelings architects are developing for their fellow creatures and the development and adoption of a feel-think, art-science design discipline.

II.4. The skin theory.

One fascinating aspect of bio-functionalism is the crucial place of skin. By definition it is at the periphery and defines entity. But also, therefore, it determines all manner of contact, and filters all relationships. The skin theory for architecture I propound emerges from this realization. In a bio-functionally conceived building, the skin, both towards the inside of a room and towards the outside, has a myriad of functions.

If we take our cue from the animal and plant world we note that through skin, simple or layered, the living things of this earth breathe, sweat, feel, radiate, attract, repel, pretend, hide, protect, absorb, insulate, reflect, transform (i.e. chlorophyll); look transparent, opaque; are light-proof, sensitive to sound and taste; it restricts, encloses, keeps watertight, airtight; it forms weapons: nails, spikes, horns, sharp scales, tails, fins, thorns; stretches and contracts; allows for movement and growth, wrinkles and is replaced. Further we note the functional variation according to position: the skin on a horse's nose is like velvet, the skin around its anus allows airtight closing up of the hole. (Note the special treatment around openings in buildings in all styles.) Enclosed by feathers, scales, bare skin, scaled skin, hair, slimy, skin, spikes, tree barks of many varieties, leaf surfaces likewise (sticky to catch insects), permanent or temporary — allowing metamorphosis.

'There is no substitute for skin', to quote Jean Ritter, my wife. I hope to make vivid why it is worthwhile to ask from building materials, when they act as skin of a building, to do very much more for us than the mere keeping out of the weather and the letting in of light. How other bio-functional properties of materials can be deployed, right to sociological potential, examples will indicate later in this book.

Just as the architect has a working drawing checklist and uses this to ensure that his drawings include all the builder needs to know, so the new checklist is intended to bring to the architect's attention the richness of opportunities every time he designs a skin for a building, outside and inside surface. Going through the list is intended to stimulate the optimum satisfaction of the bio-energetic functions through sensitive design and production of the skin.

III THE BIO-ENERGETIC CHECK LIST

III.1. Summary of categories.

The open-ended chart has been laid out in categories: each successive category adds one more function/consideration/possibility, revelant in design. The notes indicate some thoughts but the whole thing is open-ended. Specific needs for regions of the earth will vary.

Summarized list of categories for consideration of various bio-energetic functions used in the check-list.

III.3.1. Visual.
III.3.2. Tactile.
III.3.3. Atmospheric.
III.3.4. Life-supporting.
III.3.5. Energy collection.
III.3.6. Bio-social, cultural.
III.3.7. Educreational.
III.3.8. Building-mechanical-economic.
III.3.9. Maintenance, growth.

In Chapter VI examples illustrate these categories.

III.2. Visual.

(a) **Texture.** It is necessary for the eye to seek definition and differentiation for comprehension and interest. Texture provides these from close range. This is general in natural surfaces.

(b) **Pattern.** Pattern is a contrived way or composition of elements which read as one to inform. The world over repetition of some symbols and compositions suggest common function principles or bio-functional laws that govern growth and so satisfy the human eye. Pattern can orientate both outside (facades) and inside, especially in large buildings with corridors.

(c) **Form.** Form takes pattern into three dimensions and beyond mere symbols. It offers richness in decor.

(d) **Colour.** Colour effect depends on what light gets to the surface and what it reflects. Colour quality can be applied or can be indigenous, therapeutic, or depressing.

(e) **Reflection, absorbtion or giving of light.** A mirror and black matt concrete are extreme examples of the range open to the designer. Luminescence can be created through paints or through some organic substances which might be grown on walls. Light affects moods. Through its waves and stickers it can irritate or sooth.

(f) **Literal communications.** Inscriptions, from hieroglyphics, writing on wall, literal message, beyond pictorial patterns or signs and symbolism for orientation, etc., can be indigenous with building surfaces and shapes for maximum affect or harmony.

III.3. Tactile.

(a) **Texture.** The pleasures and displeasures of touching, orientation through touching. (e.g. the blind, in an institution or town centre.)

(b) **Hardness.** The pleasure of running one's fingers along smooth marble, or walking on elastic, or soft or hard surfaces. Impact with doors, surfaces, corners, furniture.

(c) **Form.** Running hand on stair rail or arm around column.

(d) **The coolness or warmth** of a surface in contrast with what's in the atmosphere, and the fingers that touch it; dryness and wetness, hot weather walking on cool surface — cool weather or warmth. Sitting with back against warm surface.

(e) **Direct energy characteristics.** We know little of these except that shock can come from touch (static electric shock in door handles can be traumatic to a child in hospital), sensuousness, orgonomic qualities.

(f) **Vibrations, energy creating,** responsiveness are very real and important possibilities . New and old fibres in clothes can create many remarkable effects to be better understood and used in all enclosing elements.

III.4. Atmospheric.

(a) **Dust creating,** surfaces abraded by wind or touch. Non-abrasive surfaces.

(b) **Dust rejecting** metallic content of surface treatment of a wall through low static current reject dust and so dirt does not settle on walls.

(c) **Dirt collection.** Surfaces can be so shaped that they not only collect dirt but with the running off of water give the dirt a pattern that emphasizes pattern and form of the wall design. Collect it strategically.

(d) **Heat** can be absorbed or reflected, minimized or maximized with many effects on exterior and interior spaces.

(e) **Precipitation.** Rain and snowfall can be modulated in its impact and can be collected for use and delight. Formation of beautiful icicles, (safeguarding against danger of them falling on people,) can be encouraged by eaves and walls. Temperature affects of rain or snow layers.

(f) **Scent creation** from walls for enjoyment and orientation, through herbs, flowers, etc.

(g) **Microclimatic effects** of extensive use of, say planted bio-walls or reinforced grass surfaces, supplanting bitumen in large areas, roof gardens, parking.

(h) **Biological effects** of lighting — (head aches from strip lighting).

III.5. Life supporting.

(a) **Plant life** supported integrally like lichen or small vegetation possibly included in concrete mix; mosses, weeds, now supported by bricks, tiles, etc.

(b) **Plants supported by soil** or medium in grooves or pockets for growth all over a wall; support of special planting for attraction of insect life; butterflies are attracted by special plants.

(c) **Support of bird life.**

(d) **Support of small animals;** lizards, fish.

(e) **Green, oxygen-creating areas.** Just as minimum of horizontal landscaping is now enforced by bye-laws, so it will be with vertical surfaces. Given cost of land, this makes sense — a new concept of landscape architecture arises: **the bio-wall.**

(f) **Roof gardens** merge into the concept.

III.6. Energy collection.

(a) **Sun** — a whole science of its own.

(b) **Windmills** of many kinds.

(c) **Rain,** for example, breaking up raindrops results in health promoting negative aeroionization (Orgone).

III.7. Bio-social.

(a) Opportunities for integral and **spontaneous participation** in creation; architect, contractor, artist, tradesman, client, users (children in schools), etc.

(b) Openness and **enclosure possibilities** in design to satisfy human needs for privacy and togetherness.

(c) Adaptability to **rational labour allocation** according to circumstances; industrialized or home-made; economy in production costs, material, and labour and power.

(d) **Unity** of style and individuality of style. Versatility for wall fittings, utensils. **Adaptability** to culture and climate.

(e) Ease of shaping **spaces for specific uses,** flexible nature.

(f) Surface regeneration — aggression creatively effective.

III.8. Educreational.

(a) **Stimulation** of imagination, feelings of belonging, orientation, delight.

(b) **Growth and change** possibilities in visual context; so that, say, aggression, inherent in our culture, is seen as a part of transforming one aesthetic concept into another, as so designed, not a matter of mere destruction (see example).

(c) **Opportunity for play:** sport, graffiti, other uses (sundials, clocks, bells, music).

(d) The **social functions** of having forms of life in the 'cities of tar and cement'. Pets for the lonely; enrichment of office and flat life. (A pigeon nesting outside an office window can make newspaper headlines, so strong is our attachment to manifestations of life, inspite of overdoes of pigeons and sparrows in many city centres!

(e) **Perma-culture** and annual crops as part of the fun and standard of life. Strawberries in every wall. I have seen in urban spaces in Californian public streets verges planted over-all with wild strawberries. Street trees, I remember with fondness from early childhood, can be cherry trees! (And I still eat the little red apples, picking them off the street trees in Canberra.)

III.9. Building, mechanical, economical

(a) **Strength,** compression and tension.

(b) **Weight** strength ratio.

(c) **Waterproof**/repellent. (Frost)

(d) **Insulation** qualities: sound, heat, radiation, light.

(e) Easy **re-use.**

(f) **Prefabrication,** central or local.

(g) **Size** of elements: labour and machinery available.

(h) More for less through design — structure is frozen movement.

III.10. Maintenance.

Costs, creative potential; renewal.

The bare character of what is considered "good" contemporary design exemplified from an "enlightened" 1973 advertisement in the Architects Journal London.

Every age has had its outstanding doors. 1973 is no exception.

Early sixteenth century
See the beautiful folding doors with their linenfold panels at Paycockes, Great Coggeshall, in Essex.

Norman
The open-heart moulded arch is the best preserved feature of this interesting mid-12th century door in the Jews' House, Lincoln.

Classical Georgian
The fine workmanship and richness of detail of this doorway at Rainham Hall, Essex, is in the typical 18th century manner.

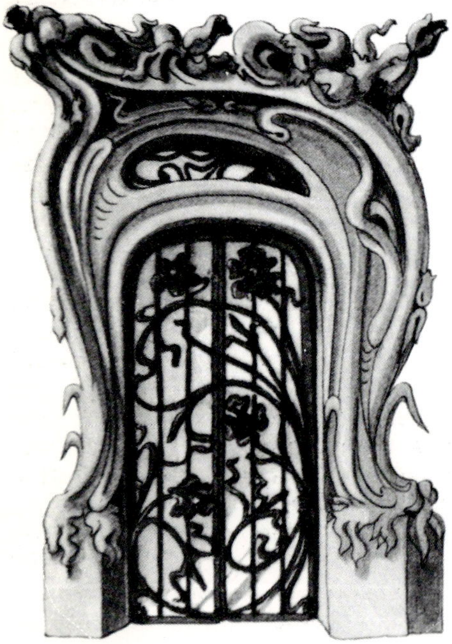

Art Nouveau
A luxuriant example of the style, dating from 1906. On your next visit to Paris, pause at No. 24 Place Felix-Faure, to admire the doorway with its full blooded curves and delicate traceries.

Crittall-Hope Luminair patio door
Constructed from extruded aluminium, white acrylic finished and factory glazed with single or double glass units. The doors slide on concealed nylon rollers. Interlocking rails, two-part stormproof sill with all-round weatherstrip of woven pile and neoprene. Immediate off-shelf delivery. An elegant door for a modern age requiring only the minimum of maintenance, offering incomparable value and fully matching the Crittall-Hope Luminair window range: doors and windows are listed overleaf.

IV THE NEED FOR SCULP-CRETE

IV.1. The lay-w/man's feelings.

Concrete could be an expression of delight. In our civilization it shows early fumblings. We have barely begun to recognize the potential let alone use it. Concrete has epitomized the insensitivity of w/man to w/man in urban design.

To give the feel of the subject to the children, at the Christmas lectures I conducted as City Architect and Planner of Perth, the capital of Western Australia, in 1965, I made my point in music. I put on two pop records: one called 'Tar and Cement' and another 'Down Town'. They echoed the ambivalence of folk to the city environment. On the one hand Petula Clark romanticizes: 'Down town, music, . . . neon signs are pretty . . . forget your troubles . . . everything's waiting for you . . . movie shows . . . little places never close . . . gentle casanova . . . lights are bright . . . waiting for you . . . you'll be all right,'

but Verdelle Smith is depressed:

'Lost in the shadows of tar and cement . . . learnt to suffer what loneliness meant . . .' and the refrain: 'Where are the meadows? — tar and cement/Where are the song birds? — tar and cement/Where is the lilac? — tar and cement/The laughter of children? — tar and cement.'

A Swedish survey showed, world-wide, that concrete was the second most disliked material substance, as I had found.

IV.2. Professional analysis.

IV.2.1. Historical context.

Firmness, commodity and delight are the time-honoured qualities essential to architecture and are still applicable. Now with concrete there is no doubt about the 'firmness'. Nor can there be any argument about 'commodity' 'a thing of use or advantage to mankind'. It's the 'delight' that we, architects, have ignored and the people in the streets have their opinions: 'matchboxes', 'rabbit hutches', 'pigeonholes', 'filing cabinets', 'cash registers', 'concrete jungles'.

The 'functional architecture' arose out of the Bauhaus movement in the 1920s which truly thought delight would come from fulfilling practical needs and functions efficiently. But they missed out subtle, human needs that emerge from the fresh bio-energetic approach. What's more, for those who had suffocated under the intolerable dust-catching burden of the opulent avalanches sliding down every architectural surface in the second half of the nineteenth century, plain surfaces would be 'delight' and the absence of ornament, beautiful. Like an orgy of food or an orgy of sex, the period of over-decoration and multi-layered adornment required a period of rest, to get over the effects of excess.

For those who have not had at least an inkling of the suffocation by ornament, it is as difficult to appreciate what I am saying as it is for the well-fed to feel hungry. Youth please note. But appetites re-emerge hence 'Art Nouveau'.

The parallel emergence of concrete as a structural and building material, with the modern, 'functional', decoration-free style has much to do with the general aversion to the material. Architects and engineers have been pleased to such an extent with the economies and structural virtuosities of reinforced concrete that they have given little attention to the people in the street grumbling under the oppressive cement monsters with their boring colours, unsympathetic textures and surfaces, demonstrating 'above all' that whereas weathering in copper, wood, and brick can give a delightful patina, what weather does to concrete is repulsive. It has been ignored, covered up, or suffered as inevitable.

But let us stay in the narrow utilitarian rut, of purely mechanistic, functionalism for a moment. Even there architects have blotted their copy-book with blotchy concrete and lack of design sensitivity. The example described is particularly poignant.

IV.2.2. None so infuriatingly blind as those who can't look out

In the early 1960s I did thorough research on deck-housing — pathways in the air to give all levels of living the social advantage of contact with passing pedestrians, and children's play alleys and spaces right outside their homes, as if living on the ground. The inescapable conclusion was the affirmation of this basic concept as correct for virtually all dense urban living. To my delight a consortium of authorities in the United Kingdom made a gargantuan effort to produce a precast system for deck-housing; streets and play areas in the air. It was built in vast quantities for many major cities. I looked forward to some joyous, satisfying feedback on those walkways and places in the air in the 1970s. But the gargantuan effort I mentioned had brought forth gigantic monsters. The little children for whom the walkways and playdecks are so vital, their only place of recreation outside their home, those little children could see only streaky concrete floors, blotchy concrete ceilings, slightly titivated concrete inner walls, and high, plain, solid, yes solid, concrete balustrades: no views. The mere slit of sky, that remained the only visible non-concrete surface was, this being industrial England, different only to the extent that it was a dirty, concrete grey instead of dirty, grey concrete. Not even tiny holes to snatch tiny views for tiny tots.

Not just one building but immense agglomerations housing tens of thousands. How many to suffer soulless confinement over the decades to come? And the professionals seemed blissfully ignorant of what they had done. When I dragged one of the responsible architects on to the scene (after all, they had been, they said, using my research), I suggested he get on his knees to have a 'child's-eye view'. * 'My God', he mumbled. That was all. 'While you are down there on your knees you might as well beg him for forgiveness', I said, in sympathetic sorrow rather than in anger, knowing that I was dealing with architects of the rarer sort, who were trying to do the right things by the people for whom they were planning. However, without models and modelscope inspection at the design stage, we will not escape such ghastly, unintentional errors. I am not casting personal aspersions. Too many well-meaning people were involved. Like the brilliant and life-positive architect in Vienna who showing me his ecstatic design for a children's home with all its nooks and crannies and spaces of fun. He too saw, once on his knees, that he had forgotten the 'child's-eye view'. He was so shocked by the bleak concrete wall blocking the view from low down,—when he thought he had specifically designed for views and children,—that he immediately called for the hammer drill, and some small apertures were pierced in the offending barrier.

IV.2.3. The hyprocritical and the hopeful.

Giving a lecture in one of the schools of architecture in London, I invited the students to look out of the window of their studio. There was the typical grey and gruesomely stained streaky concrete wall. 'Do you think', I asked, 'that the architect drew it as it is, only three or four years after completion?' They just grinned, with an idea of what I was after. To labour the point I took it further: 'But do you think the architect had a pretty good idea of what it was going to look like?' They agreed, of course. 'Then', and now I stressed the obvious, which is precisely what the schools and the professions have consistently ignored, 'is it not hypocrisy and it is not misleading a client to draw it one way, when it is known it will turn out quite different?' Architects have, with a superbly detached professional hypocrisy, continued to draw on virgin white sheets and indicated the concrete surfaces as if they were even and unmarked.

The basis of most mistakes architects have made with concrete is to assume it can be perfect. This is virtually impossible as some hundred years of buildings show. Countless builders, contractors, sub-contractors, manufacturers, and clients, not to speak of architects, have gone bankrupt or berserk trying to get perfection in a material that does not lend itself to that. The evidence of what I say is there for every eye to see in every corner of the world. At an American Concrete Institute Conference in Mexico City in 1977 I found myself in a seminar advocating that we should recognize and design with the full knowledge that concrete is aesthetically imperfectionist, like flowers and trees and natural limestone, and that, in this recognition, lies enormous postive potential for aesthetic experiment and new results which can be pleasing in many ways to many kinds of people.

One of the most refined, respected, and honoured architects of the U.S.A. gave a paper after mine detailing the incredible care required to make perfect concrete. He extolled the virtues of this approach. To counter the evidence I had given, the esteemed architect described the care he had taken over the concrete of one of the most prestigious government buildings then under construction in Washington, which, he said defiantly, had perfect concrete elements. On request the price per cubic metre of that concrete was given. With this, perhaps unwittingly, the audience and participants of that seminar realized, almost to a man, that he had just given the best possible evidence for my point of view. If **that** was the price of possibly perfect concrete then it was not meant to be perfect, and the other alternatives of imperfectionist-organic beauty must be pretty inviting, as well as generally relevant.

I believe the bio-technic, bio-energetically based styles will emerge in the next decades and will bring sense, direction, order, usefulness, and delight to our technological brilliance.

** P.E.E.R. put on a CHILD'S EYE VIEW EXHIBITION in 1959: A DOMESTIC INTERIOR 2½ times normal size.*

V SCULP-CRETE FOR ECO-ARCHITECTURE

V.1. The origins of Sculp-crete.

As a teacher of architecture in the then famous school at Nottingham, England,* in the 1950s (before it was to ruined into a University) I soon realized my degree of ignorance and that I'd better learn with the students or else. In this process and in private practice I became thoroughly aware of the limitations of concrete, and how, infuriatingly, this contrasted with its wonderful structural and plastic potential.

The problem and the need for the solutions was with me from then on. Not until about a dozen years later arose the opportunity to do something about it in Western Australia. A client said: 'Your book **Planning for Man and Motor** says "underpasses can be the brightest spots in town", can you do it?' 'Sure,' I said. 'Then I commission you to do a series of seven in my new town', said he. And I went home to dream up how to do it. And that was the beginning of Sculp-crete. I started from vague memories on the use of polystyrene as liners by Bill Mitchell. We went at it naively, pragmatically, repeating the mistakes of others, but, learning from them, our disasters turned into potential.

We saw daylight and invented a new series of creative methods and technologies. From then on we financed the various developments with sales of prototypes and preliminary production runs of flower-pots, lamp-stands, plaques, gravestones, as we advanced step by step to workable building systems, backed by patents and know-how and experience. Numberless possibilities of application tempted us. Our chief vice was curiousity not greed. We put off sober architects and builders by illustrating how clever, complex, and highly decorative sculp-crete could be. Our hopes that the few easy, sober designs we demonstrated would show that they were, of course, also possible and more economical than their counterparts already in use, were miscalculated. Again and again those who liked the idea and were attracted to the whole thing, ended up by saying: 'Of course, in this or that country it would be O.K., but you know people **here** are so conservative.' I found I could substitute almost any country's name and I would be quoting someone I had met in one of the many countries in which I had discussed these things.

So it went for some years very slowly. Tens of thousands of dollars worth of options, all over the world, but no takers. Always one on the 'Board' who had misgivings about buying the licence.

In 1977, the impact of having a 'mass-production machine' to offer as part of our package deal, made all the difference. Now we boast licences in many countries and the use of the method by notable architects and engineers is spreading. Architects who have long claimed the building industry is frustrating their creative powers are, we now find, often unable to use the freedom of choice available through Sculp-crete. They have been chained so long, that freedom is an embarrassment, it is too much. They find it annoying that to say 'I don't like it' is not valid, does not apply: for, most vital, Sculp-crete is a technique that allows each one to have what each one wants. You have what you like.

We assembled a distinguished, multidisciplinary team for a 12 hour non-stop educreational session on Sculp-crete. We designed, cast and contemplated. Even the next day the original contributions looked worth while. The applications, permutations and combinations are endless and inviting to contemplate.

In eco-architecture one is motivated to use building materials in new ways to satisfy more subtly bio-energetic needs. Sculp-crete is the result of efforts to obtain much more from one type of raw material, namely cement or concrete, by combining it with very versatile moulds. Before I describe how it is applied, its nature and evolution warrant elucidation.

The very name 'Sculp-crete' is meant to suggest that it marries and fuses sculpture and structure, art and science. It supports live functions as well as dead loads.

Concrete comes into being only when it sets **against** something. It is quite a realization that you create concrete only on casting, against something, whether you like it or not, deliberate or fortuitous. Therefore it can be deduced that a cheap material for moulds that lends itself to the easy gratification of diverse and multiple creative desires of the designer is what is needed, a material that complements the aggressive pouring of concrete on to it and vibration into it. If the whole thing sounds

* See Educreation and Ritter P., Oxford, (1979), for a detailed description.

rather sexy and fertile, its not surprising: we are talking of the close physical intimacy of materials in motion. The materials complement each other, to be fruitful, like male and female in nature. Then the pregnant possibilities abound and the method gives birth to a whole litters of ideas, each with bio-energetic potential, to fuse with material or social phenomena and again reproduce its own creative offspring!

First a short description of the three main materials combined in Sculp-crete.

V.2. Materials.

V.2.1 The plastic, setting, structural mixture.

The fundamental virtues of concrete need no elaboration in this book. It was used by the the Romans and has lasted 2000 years. Since its rebirth in the nineteenth century, concrete, (cement, water, sand, stone mixture) has been used all over the earth. The qualities in which it excells over other materials are also those that allow its development. Its plastic nature, for pouring, spraying, vibrating, shaping. Its compressive strength. It can be reinforced (with steel, fiber-glass, etc.). It can be mixed with many substances to satisfy setting time, insulation, waterproofing, colouring, and other requirements.

It can be reused as aggregate for new concrete if permanency does not appeal.

V.2.2. The expanded polystyrene form or form-liner to determine texture and shape.

The manufacture of expanded polystyrene is cheap and relatively easy. With hot steam and pressure, tiny hard balls of polystyrene are expanded and pressed together as they fill the space of a mould. The expanded beads stick together to make the required shape, most commonly a white slab about 45 cm thick, 120 cm wide and 450 cm long. (1.5ft x 4ft x 15ft) This is available all over the world.

What are its properties? And how do they prove to be complementary to the properties of concrete, enriching the relationships? This substance combines more advantages than other materials that might have been used as mould-liners:

1. It is mostly air, structured space.
2. It is very cheaply and easily manufactured, because it takes little power and only the minutest quantity of the precious oil from which it derives.
3. Easy to obtain.
4. It is sufficiency rigid to make a mould or mould-liner.
5. Insulates extremely well. This solves many curing problems found with concrete in hot or cold climates.
6. It is extremely light lending itself to, easy transport and handling.
7. It provides protection and packing to the cast article until its finish is required to be seen.
8. It has enough structural strength and rigidity to replace steel and timber moulds in precasting.
9. It is easily fixed; cut.
10. It can be easily tacked, glued, or taped to other substances.
11. It is cheap enough to make a single use economic.
12. Accepts colour or sealers on to its surface.
13. It can be sprayed with colour which then transfers most attractively and effeciently to the concrete on dissolution.
14. It can be produced in a complete range of colours.
15. It is very sensitive to heat and some solvents or high-pressure water jet.
16. It can be worked in a number of ways with phenomenal ease, variety, productivity, and economy in power, time, and labour.
17. It can be metamorphosed, from form-liner to glaze, or taken away from the cast concrete rapidly and easily.
18. When dissolved the run-off is not waste but a new material with many uses in its own right.
19. It is not toxic when burned, eaten, or dissolved. Chickens eat it with pleasure even when they have all the other food they could wish for and suffer in no way. (Eggs laid, alas, we found are not ready wrapped in insulation!)
20. It does not support flames, with additives.
21. Has an almost organic inconsistency and irregularity in the structuring of the little beads within the slab.
22. It has a texture which allows the intimate mechanical bond with concrete slurry which gives it phenomenal strength to resist concrete pressures without deforming for all practical heights (tested up to 5 m in one highly vibrated pour, to the astonishment of engineers). This penetration of slurry transforms into an artificial coral structure as can be seen when the polystyrene is dissolved and the external skeletal fragments that have formed round the polystyrene globules remain in concrete, to varying extent, to get away from the smooth grey streakiness which characterizes pours against smooth materials. It is a texture, the refinement of which arises out of a bio-chemical set of properties and relationships apt for the bio-technic era, not requiring manpower or costs. The texture has the rhythmic irregularities associated with growth.

The deeply shaped and decorative ceiling in just polystyrene is one attractive use of the material itself.

Another is that sections of polystyrene, accurately cut, will replace entirely the excrutiatingly complex steel and timber forms expensively built to cast curved bridges.

V.2.3. The solvent (if colour or glaze are required).

Pragmatic approach led us to solvents recommended by the health authorities. Subsequently having gone through the available solvents both at Utrecht University in Holland and the Institute of Technology in Perth, W.A., and with sundry chemists, we confirmed that we had the best that could be found.

1. They are standard products, cheap compared with other solvents.
2. They are easily and cheaply distilled solvents, so cyclical re-use highlights economy.
3. They are not hazardous for normal industrial or educational use with reasonable precaution against over exposure (similar to petrol, glues, paints).
4. Most important, over doses of fumes do not have a cummulative or permanent effect.
5. One solvent is not flamable — a storage and safety advantage.

It should be made quite clear that although the three materials currently chosen seem the best, we are on the look-out for better. Sculp-crete (and the patents that cover it) relates to a method, not specific materials, open-ended and with opportunity for improvement.

V.3. Sculp-crete methods: general.

The Sculp-crete method includes the following stages:

1. Preparing the form of the polystyrene mould.
2. Colouring the polystyrene mould, if colour is required, or painting on sealer if concrete penetration is to be minimal.
3. Casting (spraying, pouring, vibrating) concrete.
4. Curing of concrete. In its insulated mould the concrete is effeciently protected and can be left indefinitely.
5. Combining (or separating) polystyrene and concrete.

The description which follows gives the techniques of the patents the P.E.E.R. Institute has developed. We know that we are only scratching the surface of a vast mine of ideas. The capital equipment is simple, not expensive or elaborate — development is not expensive or difficult.

V.3.1. Shaping polystyrene moulds or liners: general points.

Independently or combined hand tools and mass production machinery, spraying and other methods can be used to fashion polystyrene. There are

Regular or irregular patterning is possible

endless possibilities. The potential is best comprehended if the illustrations in this book are carefully analysed, not for what they are, but for the potential they demonstrate.

Basic characteristics of production are: very low costs of raw material, practically no power, and unskilled labour content according to situation.

The mass-production machine rate is almost always far greater than required. The Empire State Building could be covered in three weeks. In other words an architect can come in the morning, experiment and choose a pattern that delights him. In the next hour he sees a quick-set concrete or plaster sample of it. Within the next few hours the form liners are ready to be taken to the site to be tacked or stuck to the normal-type formwork.

The process is basically so cheap that it does not add significantly as a production cost to the cost of the raw polystyrene. There is room for much profit; which is often essential to attract prospective manufacturers. The machine is simple, cheap, and adaptable.

V.3.2. Mould production: specific alternatives.

To date two major ways of working the polystyrene mould and several minor ones have been used.

(a) The mass-production cutting machine is one, the immense productivity of which is described next. (200m^2 — 400m^2 hour or 2,200 ft^2 approx.)

(b) There are easily, quickly, and cheaply produced hand tools, for rapid imprinting which can be mounted on a base, unlimited in size, to give, for large repetitive jobs, immense productivity for the most intricate type of patterns. The imprinting and cutting methods can easily be combined to challenge the imagination once again with what is the infinite number of variables that arise. These, in existence in primitive form for many years, have a new and richer place in Sculp-crete, as they are perfected to give artists wider choice, more reliability, flexibility and above all speed so that the sheer joy of carving with the movement of the wrist or arm, hitherto reserved for painting finds a new and delightful place in sculpture. The sweep of curves of the human arm in motion and the concurrent, gyrating movement of the wrist has an immediate translation into three-dimensional, permanent form. An exciting prospect and experience for those who try it. The geometric forms possible for the cutting element compounds the possibilities. The mental and emotional adjustment of working in the negative are important.

Sculpting in reverse seems the direct way of translating feeling into form and shape and then through casting into the negative mould, the sculpture.

Carving in the negative, you don't copy or translate. You feel and form. What you feel and form is then transformed into the positive so that it can be felt, viewed, and experienced by others.

I believe this is a very important characteristic of Sculp-crete, a fundamental one, basic to its nature as a new, vital and different art form, with profound, wide-reaching, and even therapeutic application.

(c) Sculpting with solvent through a spray gun and texturing and colouring with the help of stencils is a further method, again possible to combine with the others. (One 30 m x 20 m sculptured mould in two days. See illustration)

(d) Sticking things onto and into the polystyrene mould is so simply done that another set of variations offer ranges of collage or montage possibilities.

Methods, simple and effective, have been developed with which polystyrene sheets, once worked, can be bent into remarkably tight curves, making it possible to take design patterns over column surfaces, domes, irregular curves, two-way curves like hyperbolic paraboloids, etc.

The cost of polystyrene, although not great, is the greatest percentage of total material and labour costs per square metre depending naturally and normally on the depth of the pattern and so the depth of the polystyrene, and choice of pattern which eliminates waste. Undulating formliners, with the back also undulating and hollow, are strong enough. They save as much as 80 per cent of the polystyrene costs of say a 15 cm (6 in.) undulation.

V.3.3. A creative mass-production machine: 1977.

In 1977 the P.E.E.R. Institute invented and constructed a machine to demonstrate how an endless variety of textures & patterns could be produced by cutting polystyrene. The machine has been tested world wide and the potential has been aclaimed as a break-through. Resistance of acceptance by architects & construction firms relates to the conservative nature of the profession and in the industry: one of the largest form linr firms in the USA, having paid much money for options to buy the licence and for demonstrations, rejected the machine because it could not readily produce imitation brick patterns and because architects would be some five years before accepting the freedom offered by the variety of patterns the machine can produce economically.

The advantages listed below make the adoption of the technique and the machine inevitable.

(a) **enormous productivity.** A number of cutting elements working simultaneously produce 50 m^2 (550 sq.ft) of mould liner per 12 minutes. 4000 m^2 per two shift week (44,000 sq.ft.) approx.

The endless possibilities for patterns allow an individual one for each job. Once a pattern is established through creative trial and error on the machine it can be recorded in six digits and then recreated when desired at any time.

Moulds can be run more than once through the machine to have very rich results.

The deep patterns, up to 20cm (8 inches) can be cut in such a way that the emerging mold follows the wave and is only a wavy, corrugated sheet, may be one inch (2.5cm) thick. This is strong enough to cast concrete or spray fibreglass re-inforced concrete. In the latter case the finished corrugated slab can be extremely thin.
It should be noted that the waves can themselves have a wavy texture on the sloping sides.

Shallow, non-re-entrant patterns can be sprayed so that they can be physically removed from concrete after casting.

Deep, non-re-entrant patterns can be removed in part through cutting but often also just by physically removing the mould liner. Re-entrant patterns require that the mould is in part, or wholly, dissolved.

The range of pattern depth so far achieved with the machine is from 2 mm to 200 mm.

The cost of the pattern is solely related to its depth.

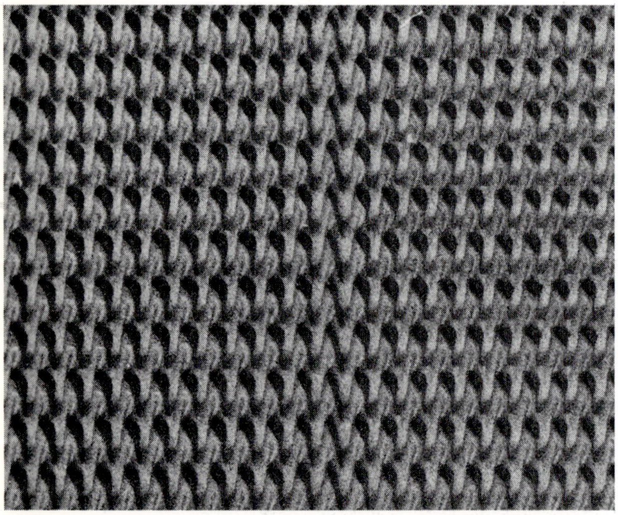

Pattern variations are extended to infinity because mould sheet can be run through the machine twice or more; either in the same direction or at right angles. Of course this reduces what is the phenomenal productivity somewhat, still leaving it within economic bounds.

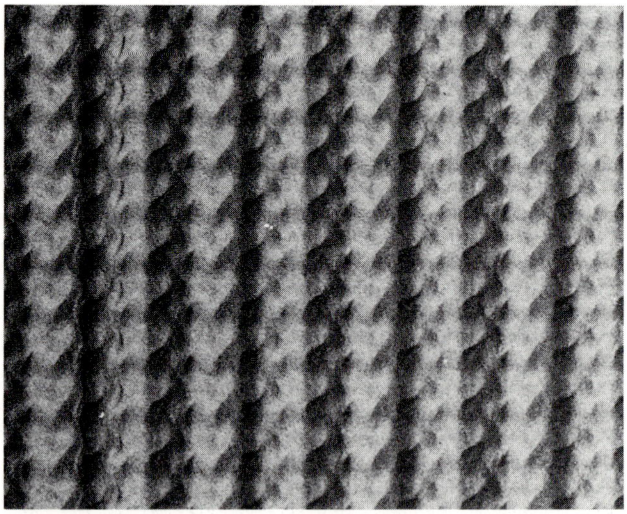

The great advantage of all non-reusable moulds is that mass-production is not held up by peeling for re-use; the mould liner provides curing insulation; it provides protection during handling time and for transport.

(b) **variety.** Without expenditure, variations of patterns are completely limitless and achieved by modifications of the vertical movement, or the horizontal movement, or the time cycle of each, or the shape of cutting elements, with the greatest of ease.

(c) Designers can, with the help of the machine, evolve designs, through instant feedback, until they are satisfied. It only takes an hour or so of 'playing about' with the variables on the machine.

(d) **no waste.** If the patterns are designed with the aim to eliminate waste it is still possible to choose from an endless series of variations.

(e) **no power worthy of note.** The machine and cutting elements take such minute amounts of power that the expense can be disregarded for purposes of economic feasibility studies.

(f) **minimum of attention and labour.** Although it is advisable to have a person in attendance to feed the machine each 12 minutes, in between there is nothing to do and the person could be otherwise engaged.

(g) **minimum capital outlay.** Capital outlay for setting up mass-production is incredibly low with minimum space requirements.

V.3.4. Mould colouring.

Recently a greatly impressed bio-chemist admired the high sophistication of the paint mixture which married so well with polystyrene and the process. It was pragmatically evolved: we had no idea! We merely knew it was easy, flexible, and economic.

Colours are sprayed or brushed on to the polystyrene moulds. This has some very specific qualities which turn out to be both obvious, and subtle advantages.

(a) Colours flow on to the concrete when the polystyrene is dissolved by the solvent. Thus, like copper glaze, it has a special character through flowing on. It merges intimately into all the nooks and recesses of concrete created by vibration when cast against polystyrene. The resulting textures the human eye finds attractive and rewarding. This is intrinsic advantage, no matter what colour or pattern. Sharp dividing lines have no place in this method, but clear colour differentiations can be obtained.

The method of spraying at an angle ensures that imprinted grooves making up patterns, are not hit and coloured. The white or grey of the natural cement then defines such projecting, cast concrete ridges (grooves in the mould become ridges, when cast). The background shows the colour. The grooves can be alternatively and specifically coloured giving another result. See Fig. 10.

(b) Colours have an intimate physical bond with the concrete and it does not flake off. If polystyrene is thoroughly washed off with solvent, colours are matt. If the dissolved polystyrene is left on then a more or less deep glaze results. The colours used are chosen from an infinite variety. Just as in tinting paint, in Sculp-crete the grey or white cement that shows through gives a subtle quality that is clearly missing from practically all coloured cement work to date.

Standard patterns can be created on 1.2m (4ft) wide units. Length is governed by the size of polystyrene available, usually about 4m or 12ft.

V.3.5. Casting.

Both normal structural concrete mixtures, white cement and sand mixtures, and fibre-glass reinforced, sprayed cement and sand mixture have been used with complete success with Sculp-crete moulds. In every case the degree of vibration necessary for strength of the concrete is adequate to fill the mould, however intricate. As the finest and strongest part of the concrete tends to segregate in the thin filligree, coral-type surface, texture, or decor, this is, thereby, given strength much greater than generally expected.

Concrete technology with all its many possibilities, mixes, additives, etc., is enhanced by the addition of the Sculp-crete finishing methods. There can be a two-stage casting process: if white and expensive mixture is required on the surface or part of it then this can be sprayed or poured as a thin layer first, on to the polystyrene liner. Subsequently the normal, much cheaper grey mixture can be added to give the structural backing.

Vertical casting, in banks or series of slabs, with liners to shape both sides of a cast element, opens up many possibilities.

Scale:

|←— 10 cm —→|

Even simple patterns have a special charm through the rough, lively, coral-like textures.

Scale:

10 cm

Rope patterns give an indication how sculpcrete can provide the fashionable effects currently sought after at much reduced cost.

Emphasis of elements, i.e. windows and doors, common in all historic styles, becomes once again feasible. The scale of this pattern can easily be ten times enlarged to form a building facade.

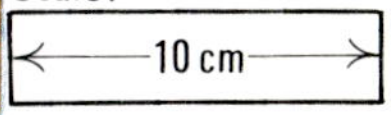

Partitions with rough textured, washable glazing, both sides and edges to match, can be cast vertically with great economy.

Scale:

10 cm

Through the Sculp-crete techniques any colour can be achieved. Remarkable effects can be gained through simply angling the colour spraying of the form liner. The diversity of effect is gained purely by turning the panel through ninety degrees. Patterns of up to 6 inch depth (15cm) can be obtained. The two pictures above are of the same sample cast with white cement.

V.3.6. Finishing stage.

General white or coloured polystyrene, through spraying or dipping in solvent, changes into a syrup-like sticky liquid which, with its colours links very intimately with the concrete previously cast and vibrated into it. The result is deep or light; coloured glazes. Thorough immersion leaves a matt-coloured surface. Peeling off polystyrene sheets moulds (separated by sealer, or removal of polystyrene by high-pressure water jet are alternatives.)

This stage can be delayed, the polystyrene having acted as a useful insulator and protector against damage and dirt in the meanwhile until the right moment has come.

(a) **Removal through solvent dipping.** In precasting work it is simple to leave elements in solvent for length of period which is regulated by (i) depth of glaze required; (ii) saturation point solvent (i.e. dissolving speed); with large slabs only the surface needs to be submerged. Distilling and recirculating can keep the dissolving strength of the solvent even and efficient.

(b) **Removal through solvent spraying.** Both precast and in situ concrete can be treated in this way. Various nozzles are used for various patterns and degrees of glaze required. In in situ work, to ensure no air pollution or nuisance whatever, the fine spray is emitted under a pyramid-shaped transparent cover. This has as its base a gutter so designed that it collects the excess solvent mixture and pipes it into a drum from which it is pumped back to be re-used until the drum is saturated. This is then changed for another drum while the first drum is distilled, immediately or later. The polystyrene should not contain water if glazes are aimed for, at the time of dissolution.

(c) **Removal through high-pressure water jet.** Where neither glaze nor colour is required and only intricately textured concrete, or the aggregate showing through, even though it follows the complex patterns, the high-pressure water jet is very effective. As the polystyrene globules are filled with water they drop to the ground and do **not** fly all over the place which is the usual and expected case when polystyrene disintegrates.

(d) **Sealer.** Where no re-entrant (cut-back) parts of patterns occur the polystyrene can by sprayed with sealer. This stops the close and intimate bond between concrete and polystyrene and allows the polystyrene to come off merely by physical force of a moderate hand-operated kind, suitable for each case. The sealer remains on the concrete as a colourless water repellant, or can be removed.

(e) Combination of the methods, particularly (b), (c), and (d) allow for variations suitable for special occasions.

V.3.7. Weathering and maintenance.

One attribute of designing in Sculp-crete is of interest. What is normally atmospheric disfigurement can be channelled so that it combines, relates, and enriches pattern and form. Maintenance is a positively conceived, dynamic part of the eco-architecture.

Where colour and glazes are used by transforming the polystyrene mould, experiment has shown that ultra-violet light (sunlight) affects this decisively. Experiments with mixes that retard this are promising. However most architects like the idea of a material that changes in weather, like wood, like stone, and for most tastes it improves visually. This is the case with Sculp-crete in the sun. As the ultra-violet creates a myriad of tiny crazes, making the polystyrene glaze more and more opaque with time, the colour beneath is protected against fading. So there is always the opportunity to recreate the bright colour and the shiny surface and transparency of the glaze by spraying with the original solvent mixture. This restores much of the original brightness, adding a quality of its own that can best be described as 'antique', relating to the look of the frescoes uncovered in Pompeii.

Another possibility is to spray the existing glaze with a solvent/colour/polystyrene mix. This not only enriches the glaze but also gives the opportunity to modulate or change colour schemes. Again, the bane of plastic finishes, the removal of the existing, does not occur. The old melts and merges with the newly applied.

On shaded walls or dull climates (north or south depending whether you build in Australia or in that 'big-headed' section of the world that somehow thinks 'north' should be on top), or inside, the finishes last indefinitely. Frost tests and water tests suggest it is highly resistant to these, because the crazing allows for expansion and contraction, while the intimate physical interconnections of the glaze which has set into the concrete, give multiple and continuous adherence.

The imperfectionist aesthetic which I regard inherent and appropriate for concrete ensures that weathering and maintenance are not critical.

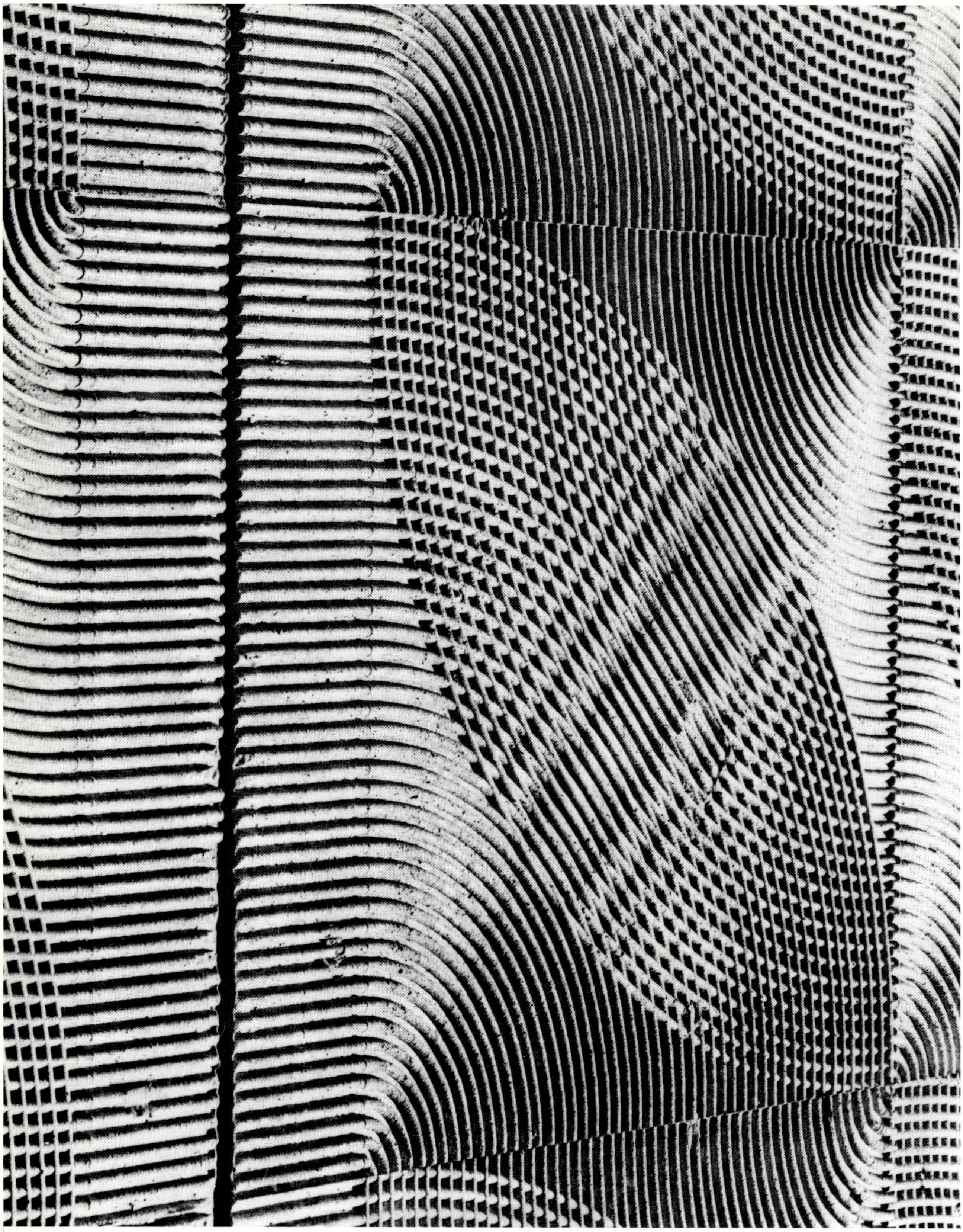

M.R.D. Bridge: The vertical expansion joins form part of the pattern which would be more effective from a distance if it had been bolder. Depending on the view quite different effects of the patterns emerge. No difficulties at the corners. Detail of cast pattern with charm of rough Sculp-crete texture. Each strand is 2cm wide. Details of the cast pattern shows an interesting alternative to the treatment of concrete surfaces.

VI Realization

Introduction

P.E.E.R. work has covered much ground but very thinly. Opportunely we have used every professional job with the full knowledge of our clients, to pioneer some aspect of the approach to improve, or extend, previous eco-architectural performance.

What follows is evidence. It is simple and direct. Check back and it will be seen how those thirty-odd attributes, which, are presented as a crude bio-functional check list for bio-technic building, are directly revelant to our own contributions.

VI.1. Examples satisfying visual needs.

Bridge Abutment Walls for Main Roads Department Government of W. Australia.

The bridge construction engineer Mr. R.A. Peters, in his article SCULP-CRETE IN BRIDGE CONSTRUCTION, Western Roads January 1977 wrote:

"The aesthetic design of concrete bridges located in urban environments presents a considerable challenge to the planner.

Site conditions largely dictate basic layout whilst financial considerations create obvious parameters within which the planner is confined.

Whilst endeavouring to balance such conflicting requirements, the planner is able to find some opportunity of creating designs in harmony with the environment by consideration of the surface finish to exposed concrete."

The result was good enough for the M.R.D. to have entered the bridge for the Australia-wide 'Concrete Structure of the Year' award.

"As an alternative to the use of conventional off form concrete procedures obtained by sand blasting or bush hammering, a recently introduced off form, concrete technique known as Sculp-crete has been adopted for the exposed surfaces to abutments and wing walls.

The results achieved on the Ewing Street Bridge demonstrate the practicable application of Sculp-crete for bridge construction, offering an interesting and individual alternative to the treatment of concrete surfaces."

Choice of pattern. The technique allows each client to design his own pattern which gives an unusual degree of choice. The M.R.D. engineers and its consultant architect inspected some sample panels: A pattern designed by P.E.E.R. was selected to be carried out with the natural concrete finish of the usual structural concrete used for abutment walls.

The dream as old as technology to let aesthetics reflect the forces in the concrete structures is so easily possible and full of promise with this technique. And the patterns that give an attractive texture, need in no way be such that they take away from the beautiful form of the curved bridge. The elegant trunks of trees are not spoiled by the attractive bark, but enhanced.

Tests. Tests carried out jointly by P.E.E.R. Institute and the M.R.D. gave the following results:

1. Normal vibrating for structural needs was adequate to fill the recesses of the patterns in the expanded polystyrene liner.
2. Normal bridge-building labour force could carry out the job.
3. Neither the pouring of the concrete nor the considerable pressure of a high wall damaged or deformed the polystyrene pattern.
4. Pencil vibrators in action bounce off the polystyrene even if applied with pressure and do not damage the polystyrene.
5. High-pressure water jet will remove most polystyrene to give a 'natural' finish to meet high M.R.D. standards.
6. Bolts inserted as part of the normal form work construction do not create problems.

Contract. The contract was based on a fixed sum and based on conditions agreed with the M.R.D. prior to signing the sub-contract with the general contractor. A sample wall was cast on site as a basis for control of the specified finish. There was a clear understanding that the costs for this pioneering project in the field of civil engineering would be considerably higher than the normal, subsequent use of Sculp-crete as main contractors charge undefined extras in case of anything new 'just in case'. Even so the Sculp-crete method was cheaper than the alternative M.R.D. finish used hither to.

Mould preparation (prior to invention of the mass-production machine). With tools, specially created for the pattern chosen, it took two persons only ten days to produce the formwork lining for the 440 m^2 (48,000 sq. ft) in the P.E.E.R. workshops. To suit the formwork units the sheets of polystyrene were produced so that the 1,200 mm (4 ft.) wide pattern covered a length of approximately 5,000 mm (16 ft) for easy fitting on site. A sealer to prevent undue penetration of concrete was brushed on in this instance. (With the mass-production machine since designed this work would only take half a day including all preparations to make similar patterns.

Formwork preparation. The abutment walls are up to 5,000 mm (16 ft.) high and the width of the formwork units, framed plywood, to be used, were approximately 3,000 mm (10 ft.). It had been agreed that the polystyrene lining would be tacked on these units as they lay horizontally on site and prior to being lifted by crane into position. This worked very quickly and very well. However, because of multiple removal and repositioning of the panels, for reasons in no way connected with the polystyrene lining, it was requested that P.E.E.R. fix some of the polystyrene lining after the formwork was in position, i.e. vertically. In spite of misgivings by P.E.E.R. with possible wind problems, the new method also worked. It was even found that to place the polystyrene sheets between the very considerable reinforcement and the formwork, was possible when variations required it. Addition or subtraction of lining patterns, when necessary, was particularly simple. All this indicated a high degree of practical adaption to the real building situation. It was recognized that considerable economies are possible as quite rough timber or damaged ply can be used to back the polystyrene lining, without of course affecting the finish.

Protection period. The concrete was poured in November 1975 and until June 1976 when, as the last operation, the polystyrene was removed, it acted as a very effective protection against damage and dirt during the casting of the deck, kerbs and parapets.

Casting. Pouring of concrete was in no way complicated by the use of the Sculp-crete lining. The glueing of the lining at the top of the formwork ensured that there would be no concrete whatever penetrating behind the lining and in places tape or a timber fillet was used for the same reason.

Removal. According to the specification the technique of high-pressure water jet was used to remove the maximum of the polystyrene physically. Contrary to fears the drenched polystyrene beads did not fly all over the place but 99 per cent was caught in plastic sheeting layed on the ground. The entire removal operation took the equivalent of three days under very unsatisfactory scaffolding conditions. Solvent finely sprayed on to the wall dissolved the remnants of polystyrene. This took seven man-hours.

Abutment walls 100m apart. Near wall shows re-inforcement in place in front of polystyrene lining tacked onto ply-formwork panels. Far wall shows polystyrene lining protecting cast wall after removal of formwork.

Pencil vibrators consolidating after pouring of concrete. Pouring in progress at far end. Typical framed formwork panel can be seen splayed out at the back.

Normal concrete poured into formwork. Reinforcement rods and formwork lining with pattern on the right by foot can be clearly seen with protection to upper edge of lining to avoid concrete getting behind lining. Satisfactory results can be achieved using conventional 30 MPa concrete mixes having a maximum aggregate sizes of 20mm (3/4in) and slumps not exceeding 75mm (3in).

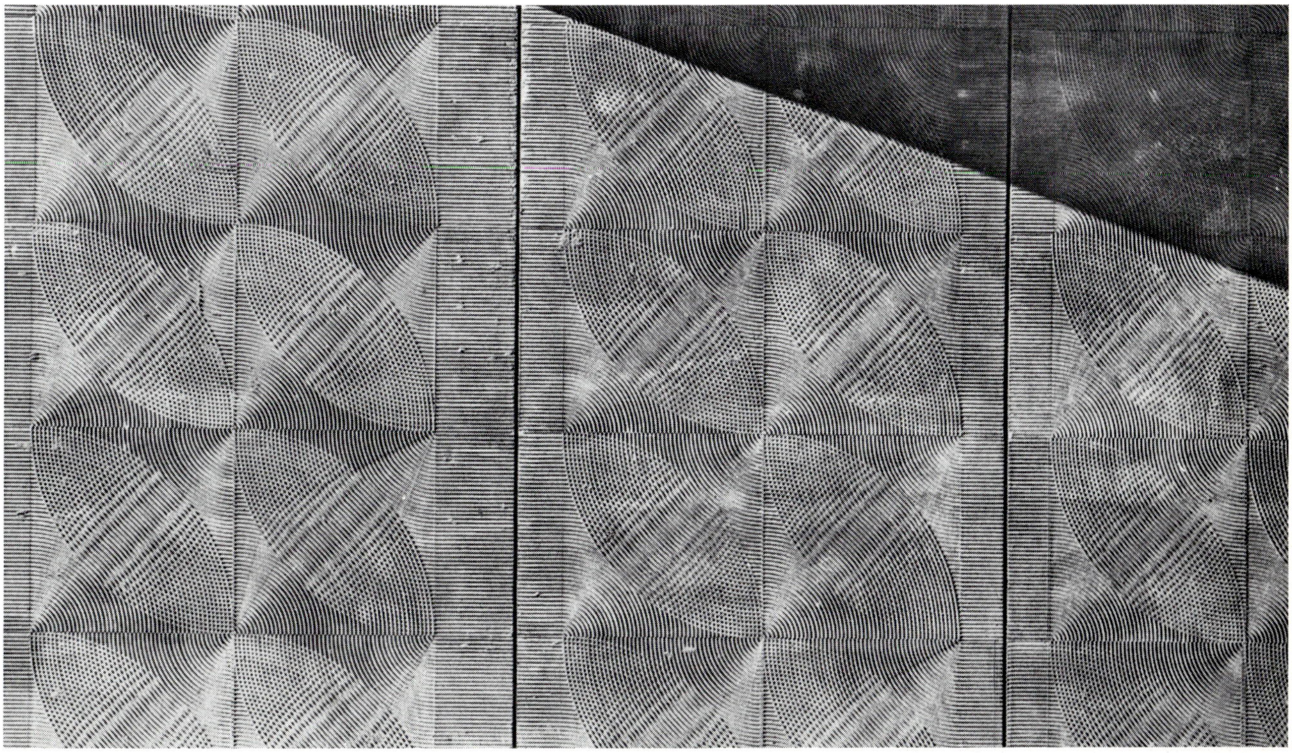

Lighting brings out the patterns under the Ewing Street Bridge. This use of Sculp-crete process for the finish to Freeway Bridge abutment walls was carried out by P.E.E.R. Institute Perth, acting as sub-contractor to the General Contractor, SABEMO, at Ewing Street, Perth, for the Main Roads Department of the Government of Western Australia. Choice of pattern should bear in mind from what distance walls are viewed. Patterns may need to be bold.

M.R.D. BRIDGE
The vertical expansion joins from part of the pattern which would be more effective from a distance if it had been bolder. Depending on the view quite different effects of the patterns emerge. No difficulties at the corners.

VI.2. Examples satisfying visual, acoustic, and tactile needs.

To understand the tactile function it is a help to close one's eyes for a while and imagine one is blind. Orientation as one moves about a building, for example a hospital, a park, a city, is then largely by touch and sound. The much-needed help for the orientation of the blind, the patterning and texturing of pavements and walls for orientation purposes and indeed their pleasure can give tactile, acoustic, and visual variety and delight to the person with sight also.

Braille signs, both on floors and walls become quite simple.

The noises of the wind and possibly water gurgling, spraying, dripping, rushing, tinkling, roaring, pouring down. Water is an important tool for the designer. A building material easily shaped can influence the sound of quite small streams of water in urban areas. Through trial and error the Sculp-crete shapes can give a landscape architect just the effects he is after.

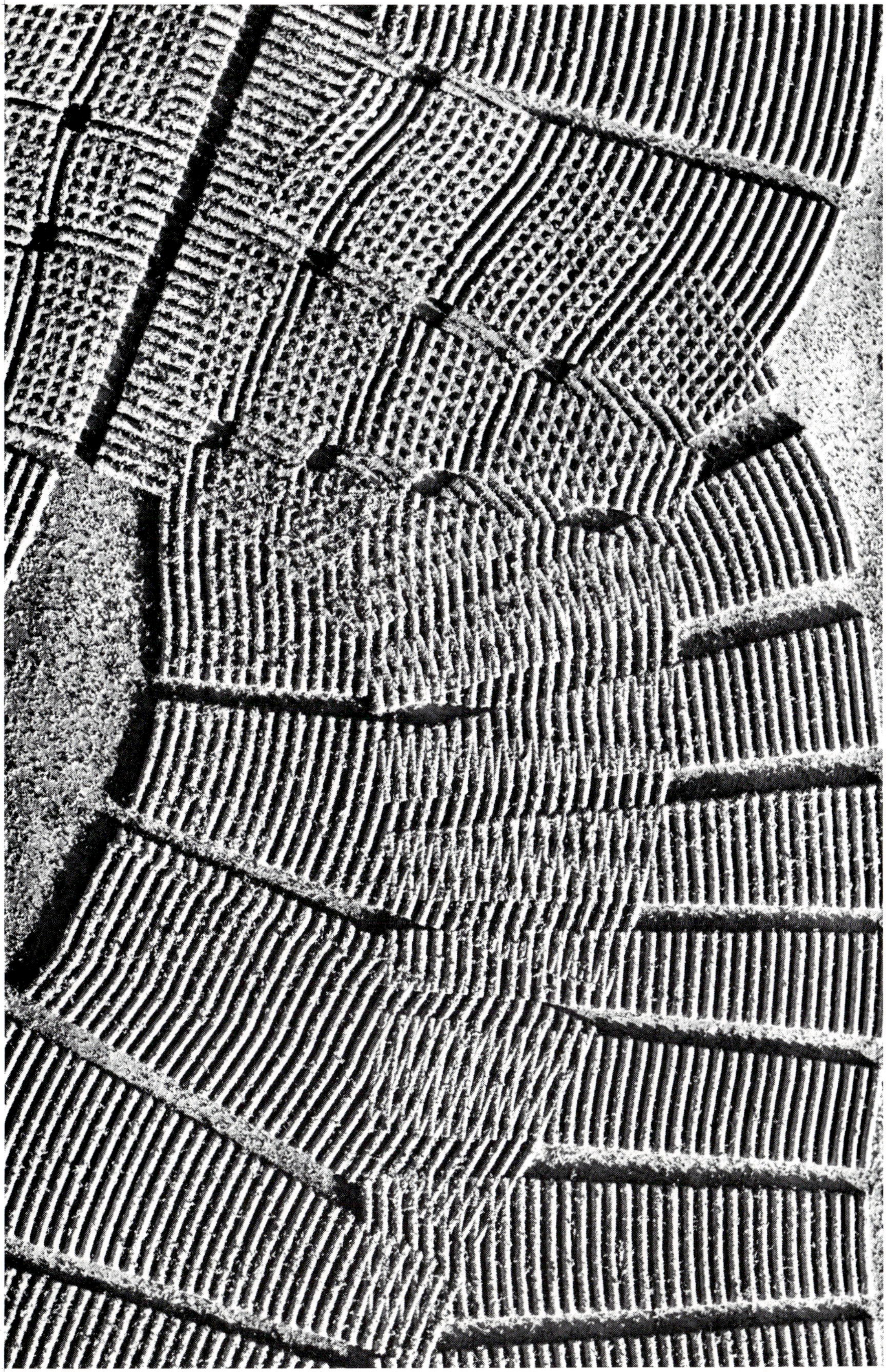

An architect or artist could, with such bold 40 cm (16") wide strokes of a multi-groved heat tool, design and produce, directly onto the mould liner, decorative treatment in scale with his building. The spontaneous feeling for his building and what he wants it to express can, in this way, appear on the 'skin', either cast in place, or pre-cast in elements that bring the pattern together when fixed on site.

Scale:

10 cm

To orientate by touch of destinctive patterns can be a pleasurable and useful function of the skin of a building.

Scale:

10 cm

The textures chosen in Sculp-crete can follow corners or curved surfaces round building shapes.

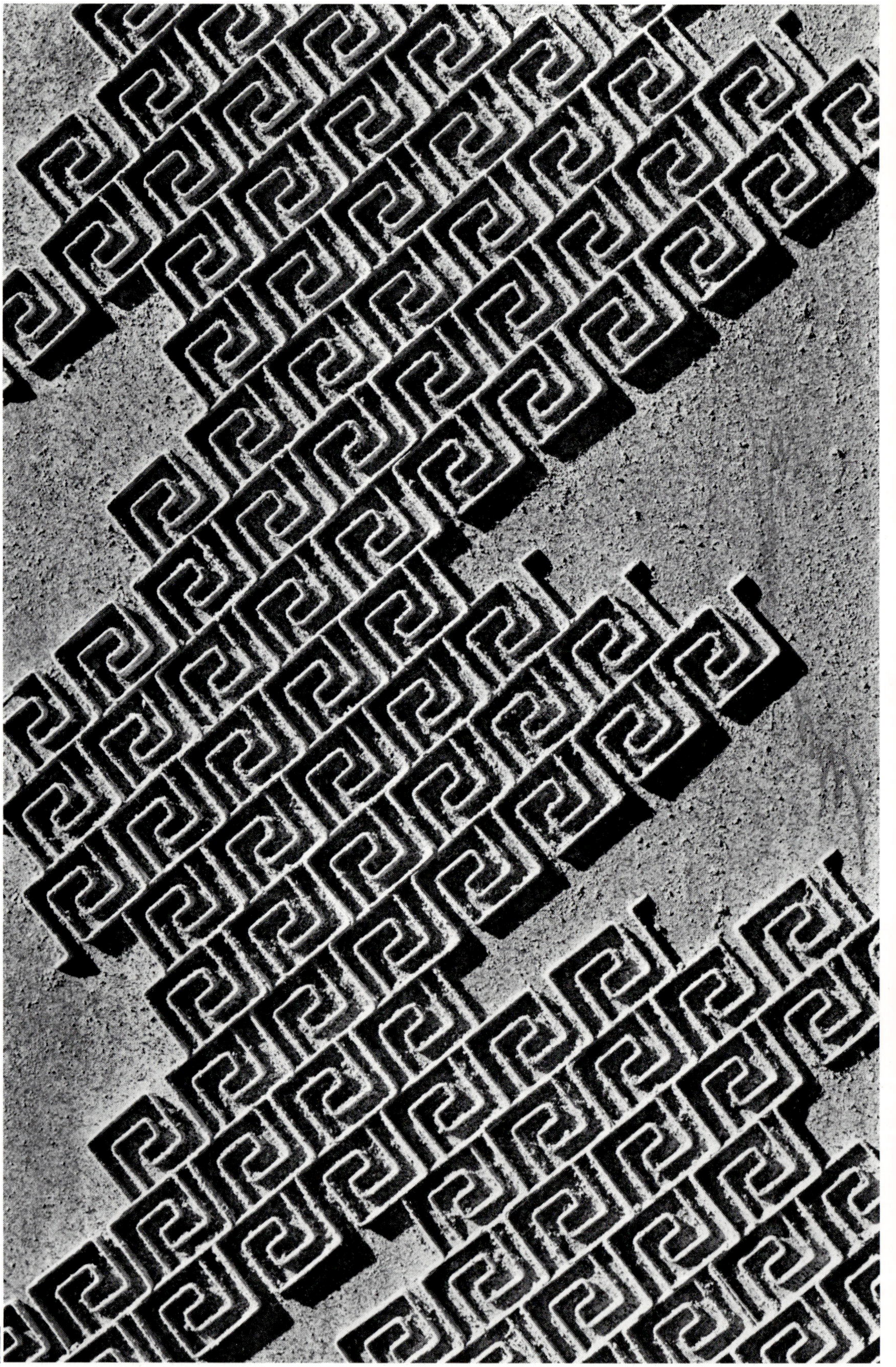

Scale:

10 cm

Tools and elements for tools can be created to suit individual design in minutes, by hand, easily and cheaply. In fact the cost of variation is virtually nil.

Tape design on moulds and subsequent light solvent spray creates patterns through difference in texture.

Scale:

10 cm

Gay abandon within economic reach of every builder and architect. It will soon become inconceivable to think of concrete without a specific character, scale, depth of pattern and texture in mind. Just "concrete" will cease to exist.

Scale:

10 cm

Appropriate symbols can be incorporated in the structural casting of a building and superimposed on all-over textures.

Scale:

10 cm

Both those with sight, and the blind, could follow the "daisy route" to destinations in towns or official buildings.

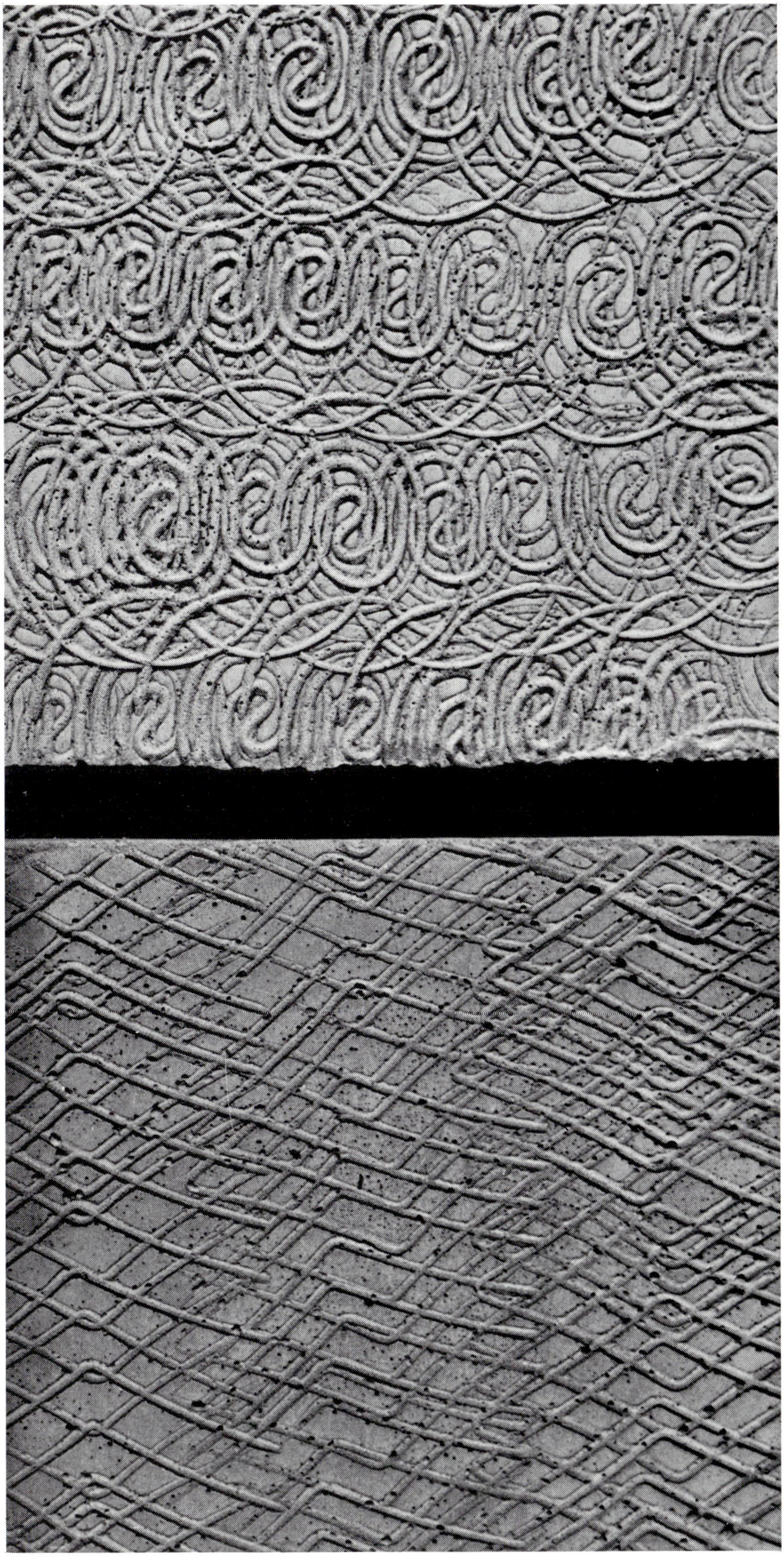

Scale:
10 cm

Paving can be designed to harmonize and have a unity with the other elements of the building it abuts. At the same time an anti-slip function is created.

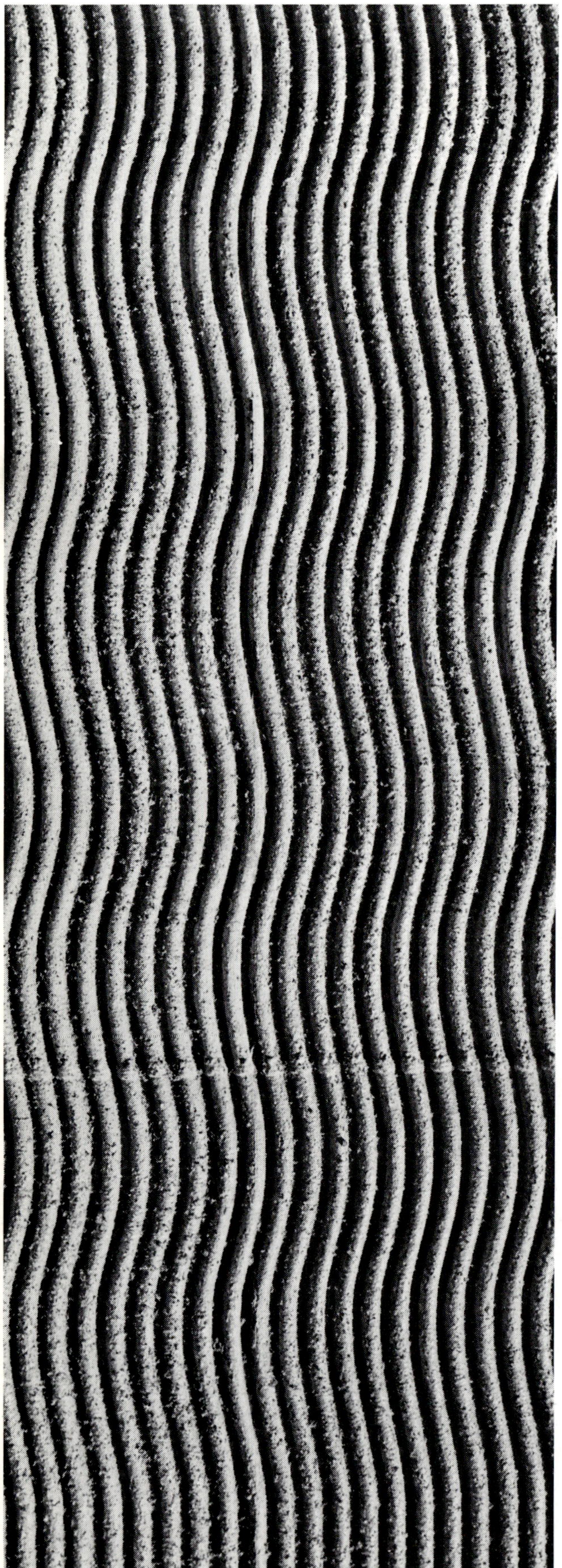

VI.3. Examples satisfying visual, tactile, and atmospheric needs.

The skin of buildings and the urban landscape generally have multiple effects on the atmosphere which were ignored hitherto but which will be considered creatively to an increasing extent as the bio-technic era comes of age. I take the definition of 'atmospheric' very simply as the air we breathe with our lungs and the pores of our skin. Now the fascinating thing about skin is that it is at the meeting point of relationships. What happens there affects outside and inside of each entity, as well as the entities as a whole. I touch ice. My skin does something externally; I feel cold internally (I react). The ice having been touched by my hot hand, melts.

A building hit by sunshine can reflect and absorb in many ways. In many climates this has many sensitive indications for design of building skin.

A building hit by sound can deflect, slow down, or make noises.

A building hit by rain can create delightful water patterns, water noises; the atmosphere comes in. As splitting droplets create negative ions, which have a powerful health-giving effect, this will be a major concern in building.

A building surface can collect dust and let it be blown off again into air, and yet we now have the option of electrostatic through the metallic content of paints or glazes on the surface of walls, to reject dust and dirt. Therefore the surface stays clean.

The ease of determining the nature of the skin of the building, afforded by Sculp-crete methods, invites a far more general consideration of the possibilities. It is quite different when good ideas are pie in the sky. Now ideas can be tried and realized with economy and ease.

Wind, water and dirt can be guided to accentuate the function of the skin of a building.

Scale:

|←— 10 cm —→|

Sculpture by Ralph Hibble.
Waterfalls on such forms maximize droplet shatter. The refinement of detail possible through biochemical actions gives the technological counterpart to growth — the concrete, coral-like, has formed a matrix of external skeletons around the polystyrene globules, which are left standing when the polystyrene is dissolved.

Sculptures monolithic and fantastic can be created to form a unit as the monolitnic piece of concrete above or created in a linear form at a much larger scale to give, for example a security barrier which does not stop air-flow and which can combine with vegetation.

Scale:

10 cm

To allow cooling breezes to blow through an urban environment is an ancient idea. Sculpcrete can not only give grace and individuality to this, but also on grills (see above) a different pattern can face each way thus giving a chance to change scale. For example balcony rails can have bold scale to the street and refined scale towards the home.

VI.4. Examples designed to satisfy visual, tactile, atmospheric, and life-supporting needs.

The establishment of settlements in human history has usually been associated with 'clearing'. Bush, forests, swamps, vegetation in general have been burnt, filled, cleared to start with. Tigers, trees, snakes, and mosquitoes seemed the number one enemies. The truth of this is self-evident: forty men were eaten by two lions when the railway from Mombasa was built (I just heard on the radio). The city of which I am a councillor commemorates its founding, exactly 150 years ago, by a brass plaque telling of the felling of the first tree

As urban areas have grown hard and huge and inhospitable in turn, with peaks of concrete, valleys of bitumen, and rivers of rubbish, and winds of stale, poisoned air, the awareness of the importance of life and its supporting systems and ecological balances became obvious. Today it has become the most universally recognised value. The task of cleaning up the earth lies before us. And cleaning up usually takes a little longer than messing up. I'd say the next millenium.

One of the earliest and simplest remedies was the introduction of greenery. Bring back some of that nature that was cleared away. Apart from the very basic exchange of oxygen and carbon dioxide between man and plants, there are matters of scent, the quality of air that it is a pleasure to breathe, delicious, and the visual delights that spring from butterflies, birds, and other forms of life as well as a host of more subtle energetic relationships which we have only glimpsed and not fully understood at this early bio-technic stage.

Now the hard facts are that land costs and the whole urban system tends towards high and dense building. This has left cities very short of planted areas, particularly in city centres where the high density of people make it especially desirable. Just as in residential layout it is now obligatory, the world over, to provide a certain area of open space, so I believe the time is very close when the need for ecological equilibrium will become statutorily imposed with office buildings in the heart of towns, and so vertical landscaping will become an additional and special professional skill.

It makes simple sense; where there is a premium on horizontal and ground space, vertical surfaces and roofs should be used to support greenery and life. My concept of the 'bio-wall' which has emerged from this, lends itself to endless physical expressions and variations. It is the economic feasibility, with Sculp-crete, that is so exciting. Here is an integral way which brings delight to hard architecture. I have watched it on the faces of people of many lands in both hemispheres as they are shown the photographs of concept. Plant pots are so often added to architecture that the desires are obvious. The life-supporting bio-wall will develop as landscape architects and engineers maximize the possibilities and the appropriate vegetation (perhaps permaculture), insects, and birds. With the fitting growth media, weather resistance, and moisture supply, it will give the kind of richness of sweet experiences that the city centres lack. Skyscrapers can waft brush-soft foliage against the sky. Tree-lined skylines will be 'in'.

The architectural, stepped back, terracing revived so brilliantly by Faller and Schroeder from Stuttgart and which has spread across the world so rapidly, has, as one of its chief delights, and raison d'etre, the greater enjoyment of plants, climate, and total environmental delights.

'Permaculture' and other ways of growing food efficiently and with taste in every sense, in small quantities and in small areas are developing world wide, once again. Formerly economic necessity. Now often living delight and delight of living.

As Colin Buchanan coined the word 'traffic architecture' to indicate the very intimate connection between building and transport, so I have coined the word 'eco-architecture' to make designers aware that this consideration is relevant to all aspects of urban environment. It enters when you choose and shape the very materials for your walls and floors, the all-important skins. The bio-wall is a concept capable of great development in depth and variety both figuratively and literally. Bio-architecture is an alternate name.

'Garden Cities', 'Cite Jardin', Hanging Gardens of Babylon — the idea has been there all the time, when and where the urban hardness spread. The princes have had their palace parks, emperors their hanging gardens, kings their conservatoria, dukes their pets. The status of the office boss is endorsed by the size of his pot plant even though it may merge in the office landscape, engulfing whole floors with greenery even in private and bureaucratic strongholds.

In a work of supreme genius, the little known or appreciated deck-housing scheme in Rotterdam, the Gallerij Bouw in Spangen an epoch-making landmark of city housing, constructed in 1920, has the soil containers built into the concrete railings of the street in the air, as an integral part of the architecture.

Architects lag behind. People are asking for these things. They had to wait a generation for awareness that a reasonably efficient kitchen in a home is essential. Access for the disabled is still being ignored very widely. All these and much more will come about, **par excellence** in the third millenium. And a thousand years is just ten one hundred year old people laid end to end. Not much.

Scale:

10 cm

IV.4.1. Bio-walls.

One version of the bio-wall: a continuous protected groove for planting. Up-side-down this pattern makes a sun-louvre for tropical use. Continuous planting has the decisive advantage over creepers that death does not create a problem of cleaning walls.

VI.4.2. Mini-growth on bio-walls.

Mosses and tiny scale plants, self sown or introduced to concrete walls by judicious choice, humanise surfaces in a way that is enchanting. Sculp-crete surfaces can maximise this natural phenomenon and the process can be speeded up. In areas with frost the matter requires a special attention. It is true that nature can destroy the efficiency of building materials and this must be born in mind. However this requires the solution of specific problem's rather than abandonment of the idea of bio-walls.

This piece of bio-wall is only five years old and yet it has the richness of an ancient wall. Pattern, texture, nature combine to give this colourful and sympathetic effect to a "concrete" wall.

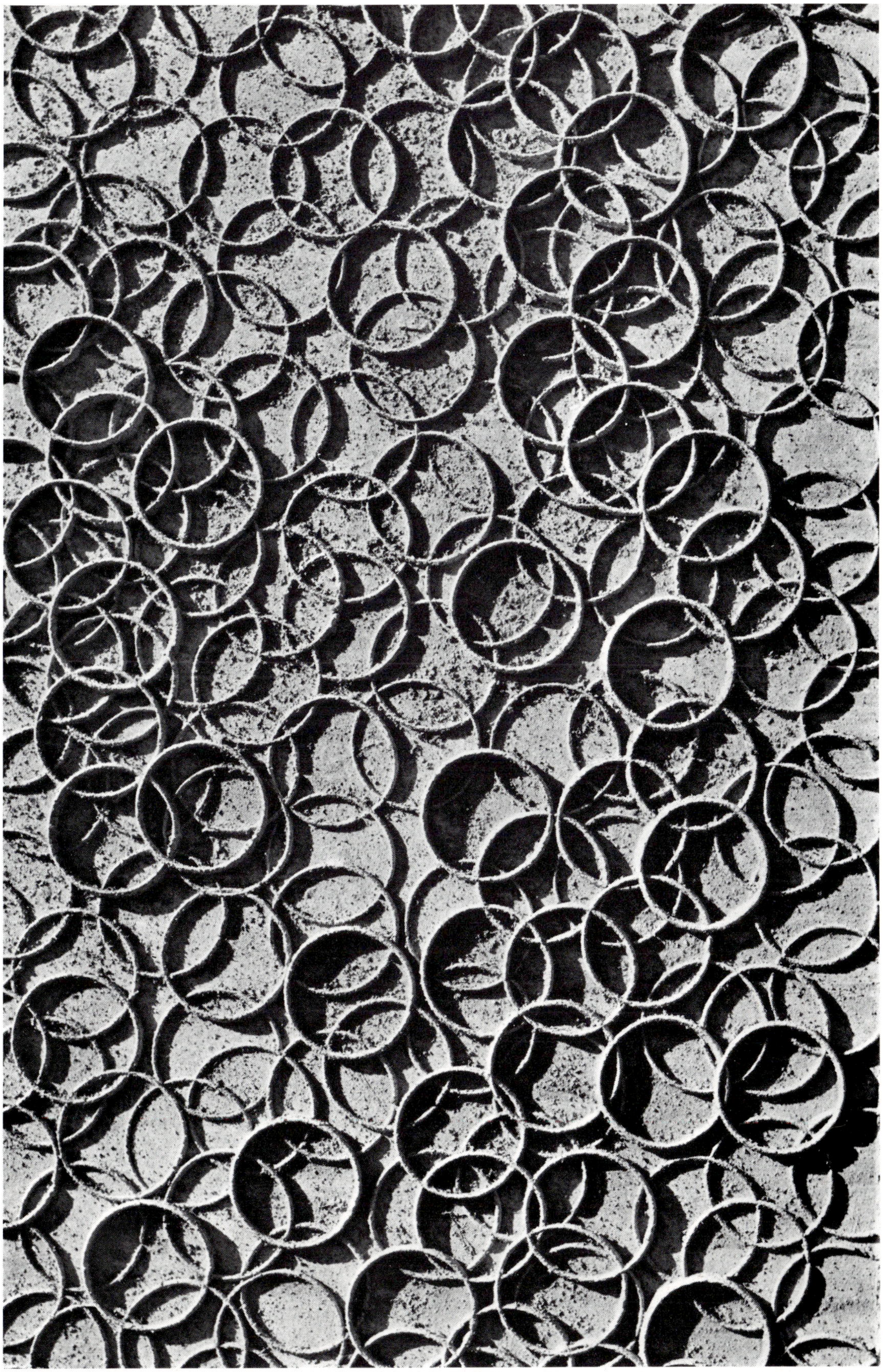

Scale:

10 cm

Shortage of horizontal areas for plants to give ecological balance in cities gives rise to the concept of the 'bio-wall'. Research and common sense are required to pick the right growth medium and plants for each climatic occasion. Drainage and reticulation are easily provided.

Scale:
10 cm

No surface in nature has as little texture or pattern as the lifeless geometry of modern architecture.

VI.4.3. Bio-floors

Ritterings. Roof gardens have been costly considerations. A great depth of soil has been regarded as necessary for sustaining growth, even excluding trees. A new approach to reinforcing ground, particularly suitable on roof gardens, has been evolved and used by our P.E.E.R. Institutes 'Ritterings'. Rings, polyvinyl, very light, very strong, very simple as well as cheap, formed into mats. It should become a matter of course to plant on roofs in soil 2-5 cm (1-2 in) deep. The view down, on to roofs, an ever more common outlook, is, hardly anywhere considered seriously or systematically. They should have the pleasant common denominator of being landscaped areas. This works also in areas of car parking on the ground and on roofs. The vehicle wheels are supported above the soil level so that, except for excessive use, grass, etc., grows where now vehicles stand on bitumen or concrete at any level. This works in all climates as it is easy to work out and provide the right plants, the right moisture for best growth, and minimum maintenance is required. Natural drainage in soils like sand conserves waters. This is a way of 'humanizing concrete' by replacing it.

This simple system works because traffic destroys lawn through undue compression of the soil and tearing out of the roots. The network of rings stops both these actions.

The intricate and flexible combination of netting, roots, rings, and grass, consolidated with soil, has remarkable strength in compression, tension and sheer. Adapts very well to the minor undulations.

Normally the rings are not seen and the lawn is indistinguishable from lawns without rings.

The inserton of VI.4.2. at this point is determined by back to back colour plates.

1. *Unroll ring mats then unroll turf and vibrator roll.*
2. *No rings visible on finished lawn parking.*
3. *The ugliness avoided.*
4. *Mats are very light.*
5. *37 ton crane running over bare mat on sand. NO BREAKAGE.*

The Government School architects have chosen Sculp-crete to allow children to join in production and to save on brick cost.

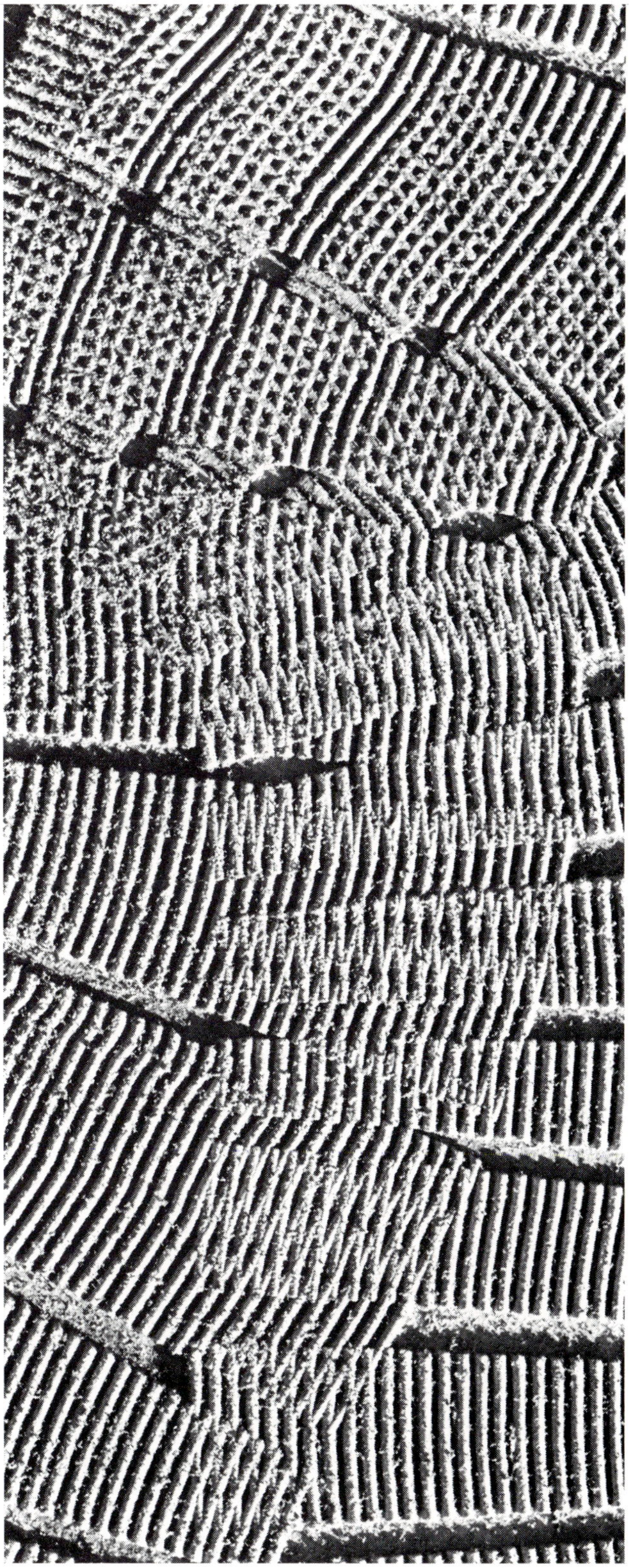

Immense, free hand basrelief designs by architects for multi storey surfaces are now economically feasible. With tools up to one meter wide they imprint to the varying depths required with the sweep of the arm of the artist, directly onto the mould, laid on the ground. For bridges etc the stress lines could be given external expression. Pre-casting or on site casting is possible.

VI.5. Examples designed to perform visual, tactile, atmospheric, life-supporting, and social use functions.

I know that humanity will for a long time suffer from 'emotil', and that one symptom of this is aggression, that needs to come out, particularly in children. Thus, when I was asked to design seven underpasses so that they were a positive addition to the environment as part of a footpath system in a pioneering Radburn housing scheme, carried out by my friend George Clarke in Perth, I realized that aggression was a crucial point. Unless one had in mind that kids needed to let off aggression it was perverse to create beauty in the underpasses. Even if desirable, which it is not, it is impossible to stop kids from letting off aggression in such places as underpasses and towards such things as walls and their protuberances, patterns, etc. Thus the design incorporates the idea of the violent attacks on it. They are anticipated and planned for, positively.

The result and the feedback, seven years later show that this can be done and has wide implications. A sculpture can live through phases of life — like any other living thing. The Venus we adore has lost her arm, Socrates is minus penis, the saints in British cathedrals were robbed of their heads by Cromwell's men, and the Acropolis is visited by thousands of admiring architects, though it is blown to bits. Who will tell me that it is not possible to plan for this kind of thing in design, particularly in a minor way? Seriously, a dry mount imperfectionist aesthete is appropriate to nature and to a dynamic eco-architecture.

VI.5.1. First educreational application — underpasses.

At Rockingham Park, Perth, W.A. 70 m^2 (672 sq. ft.) of base relief sculpture, made for seven pedestrian underpasses.

The concrete was cast by Rocla Pipes Co. at Perth. 'Sculp-crete' work and play sculpture ideas by Paul Ritter.

A 90 cm (3 ft) high sculpture panel is cast, integrally with each precast structural section, giving a continuous base relief, 10 m (32 ft) long, in each underpass.

Continuous lighting is incorporated on the opposite side and seats are placed on this side also.

From dull tunnels the underpasses are transformed into shady, dry, light, and attractive playspaces or 'clubrooms' for the children in a climate that needs shade in summer and shelter from heavy rain in winter.

Brief for underpass sculptures.

1. To invite children's interest.
2. To be vandal-proof.
3. To be attractive to adults.
4. To be economical.

Solution: 'Instant participation sculpture'

The criteria, according to developers, builders, artists, and child-psychologists have been met:

1. Rough textures invite the children to 'pick' and 'dig' as they like to do. This is intended. There are some 'treasures' hidden (marbles, etc.) under very thin layers. All this does not spoil the design but helps to give it the intended changing character.

2. The hard nature of finish and concrete make the works as vandal-proof as is possible. To be able to do many energetic things with the sculpture lessens the desires for misuse.

3. The designs have a basic artistic intent. The quality of the design texture and finishes give beauty at various levels of appreciation.

4. The incorporation of the design and finishes in the structural precast formwork reduces expenditure dramatically.

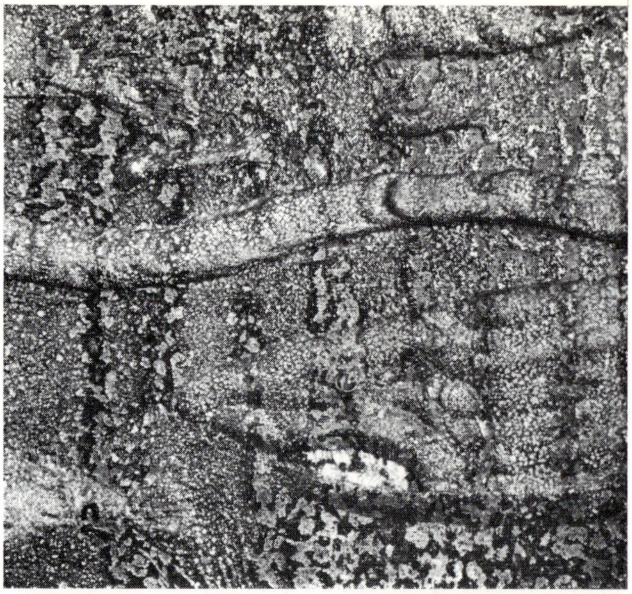

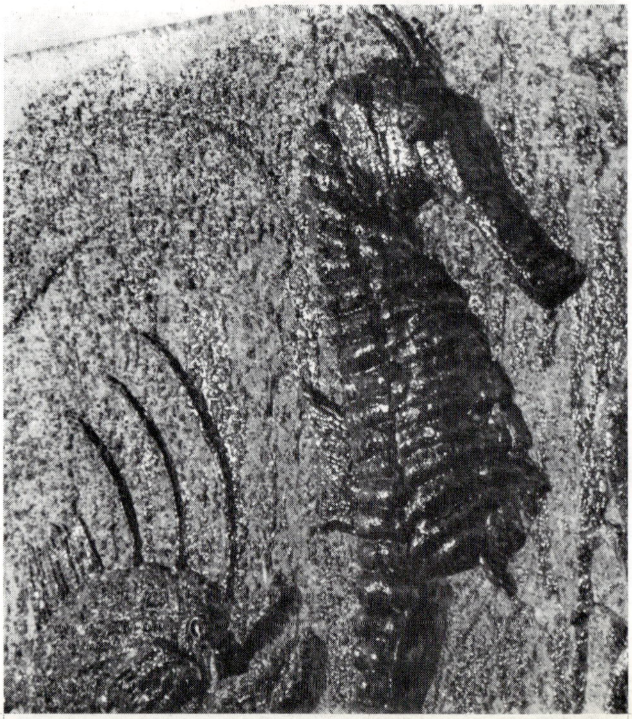

UNDERPASS 1
Underwater scene: Children imagine themselves heroes in dangerous situations. The shark, looking ferocious and with its teeth representing danger, tempts the child, I remember from myself to dare to put a finger in the shark's mouth, quickly, to see 'if anything happens.' Result: feeling of heroism and reassurance that it's all right. The chase of the small fish by big ones, the squids, and sea serpents give food for the imagination.

UNDERPASS 2
Tech fun: Ball bearings in runs or coils of pipe are imprisoned in the concrete panels with various distinctive and decorative patterns. They can be reached with pencils or 6m nails, and pushed and set off with great velocity. This allows letting off aggression quite harmlessly as the balls crash and clash with ferocity. Then the child can settle down to any of a number of games of skill, not only competitive, but games of co-operation, with couples acting in concert. Just for fun of the smaller children, balls can be rolled in and out, up and down.

UNDERPASS 3
Birds of Paradise: The glamour of the shining surfaces lit up by the coloured lights opposite, stir the imagination of the children. The occasional bit of glass should come out and will be treasured like a jewel: by the children this will not ruin the effect as the concrete is completely smooth where the glass may come out and has a continuing sculptural effect. Eyes of shining beads or marbles give realism.

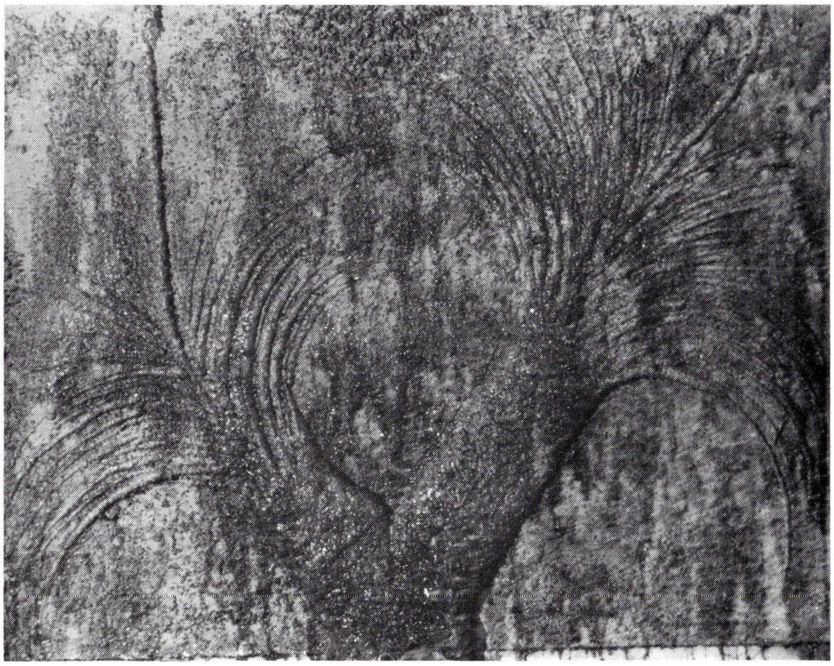

UNDERPASS 4
In the bag: A scene of kangaroos and blackboys. The pockets of the kangaroos, and in two cases of the joeys in the kangaroos' pockets, invite games of lobbing little stones into them, as a game of skill. The joey's pocket is obviously the 'bull's-eye'. The life size of the kangaroos give excitement to the scene, and the little joeys will attract the very young.

UNDERPASS 5
'A Keep-em-Rolling': Wheels of many kinds are mounted in patterns. Children can make them go round quite fast and the fascination of never having any of the wheels still, or half the wheels still, can give amusement for single children or for groups in organized games or informal fun. The turning of this fitting can also be pretend technical equipment for imaginative ships to the moon, and the like. Ball bearing races run in various tracks so that children can play 'trains', 'train crashes' or more sophisticated scoring games.

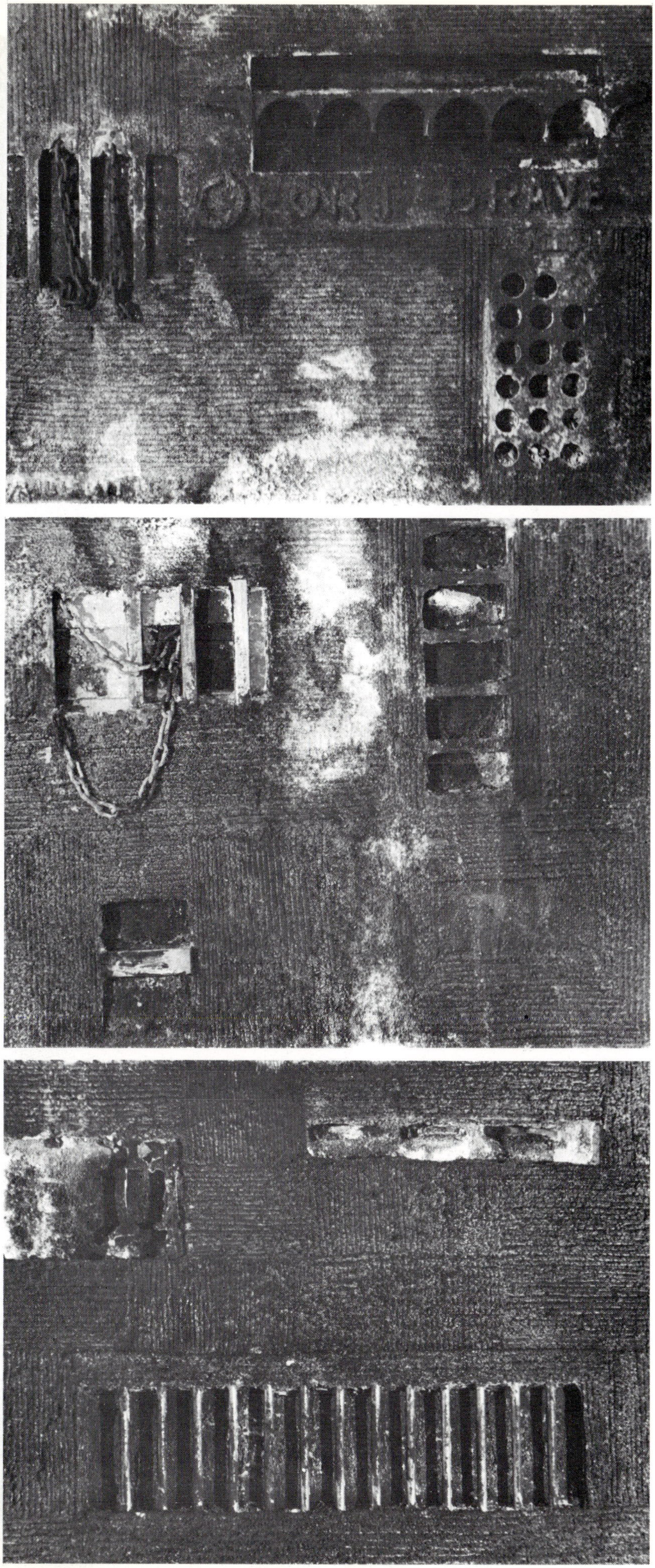

UNDERPASS 6
'Fort Brave': Series of iron windows in walls, with chains and buried bullets. The bars in the windows give a handhold for the children while projecting shelves give footholds. This allows children to climb through the underpass without touching the floor, a favourite game. It also gives children the chance to run a stick along the bars making the characteristic clicking noise liked by children. The most successful of all the designs in use.

UNDERPASS 7

Fish and chips: Large fish with mouths stated so that they act as receptacles for little 'chips' of stone to be lobbed into them be the children. It's a simple scoring game. The large ferocious and very varied fishes and sea horses, with their very realistic eyes are designed to stimulate the imagination.

They did. Little boys lit harmless, tiny fires in their mouths. All the blackened mouths were very attractive while they lasted. Participation indeed. Fun and no harm.

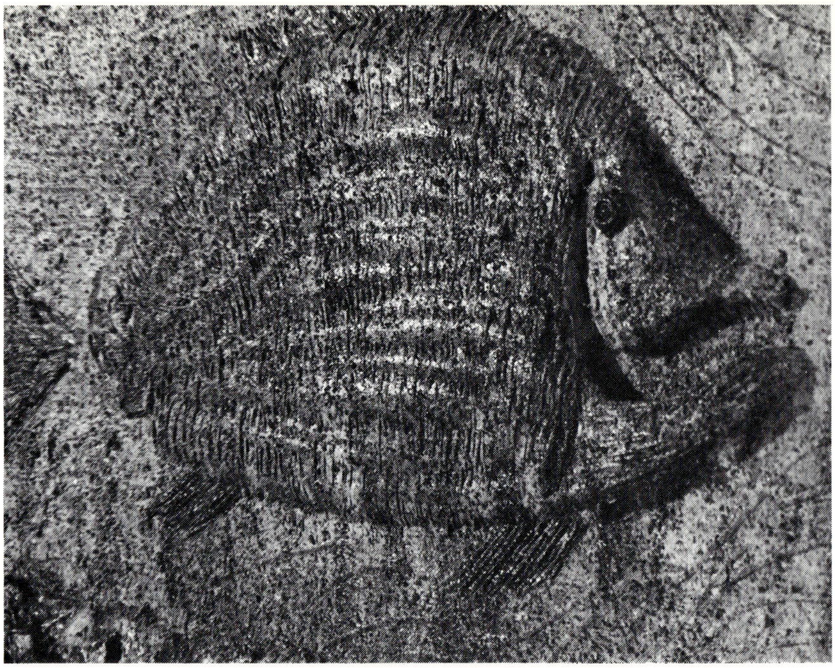

FEEDBACK!

After seven years the underpasses are operating much as planned. The evidence of thousands of hours of childrens activity is right there. Seats were never put in and cycles are coming through a bit fast at times but can be slowed down. The surfaces are picked off as anticipated and the glass mosaics below are beginning to appear, with mystery and fascination and invite more participation. Marble eyes have been bashed, or knocked out whole, to leave an interesting rough glass texture, or a shiny smooth concrete hollow. The attraction of some of the ball-bearings was too great: hammers have dislodged them.

Even where games have been distorted the remains give new scope for play. It is so much better than theusual walls with just graffiti or smooth patterned tiles. The all-over graffiti on areas of concrete not covered by Sculp-crete fit in with the non-perfectionalist aesthetic of Sculp-crete. What would, elsewhere, be seen as damage, fits in as growth, participation, and natural aging of man made and man used environment.

The four points of the Brief set out on page 65 have been met.

VI.5.2. Educreational kindergarten experimental design, Rockingham Park Kindergarten, Shire of Rockingham, Western Australia, 1970.

Designed by Paul Ritter, Ralph Hibble, assistance from Mark White, and Wally Greenham and the Planned Environment and Educreation Research (P.E.E.R.) Institute, 1969 for Rockingham Park Pty. Ltd.

Location. The site is in the first section of the New Town of Rockingham, planned by Clarke-Gazzard Partners, for some ten thousand dwellings with traffic segregation on separate path and road systems.

The kindergarten site is along side a delightful major path of the first residential area, which also leads to the primary school, new town shopping centre, and playgrounds.

The pedestrian entrance is from the path system. It emerges from one of the seven 'participation sculptured' and specially side-lit, underpasses.

Brief and aims. The developer asked for a 'showpiece', costing no more than the normal kindergarten.

This coincided with the Institute's educreational outlook.

Ideas and concepts. All the requirements could be met in a pavilion like building form: six hexagonal shapes, forming one-seventh at their centre, all unified under one, simple roof, composed of three facets, each a flat hyperbolic paraboloid, pitched upward to the centre and crowned with a 9 m (30 ft) high, slim spire.

This created an appealing central space with almost 360 degree vision of the many activities taking place in the part-hexagonal rooms at the periphery, and further into the **six verandahs,** one outside each room.

In this way the activity of each place could be reflected in the treatment and character of that part of the building, outside and inside.

External Appearance. The spire with weathercock, covered in shiny, coloured glass and orroglass, some 15 m (50 above) ground level, is intended to mark the importance of the young child in society, as well as show a sense of fun. In the sunlight the spire shimmers and at night its internal lighting will glow with a warm, burnt orange colour, and will be visible from afar on the flat landscape, giving a dominant feature to a sea of low buildings.

There are welcoming verandahs on every side. The unity of the exterior is created by a continuous frieze of a bass-relief of happy children's faces, with a natural concrete pilaster effect each side of the eighteen uniform, square, sliding, aluminium windows. The natural-brown, timber-framed, spacer panels with the six doors, leading in from each verandah, add warmth of colour.

Lettering, giving the name and nature of the building is cast into the structure. The inscription, 'we love you', is to summarise the educreational and design attitudes.

Participation of the children, even in the day-to-day external appearance of the building, became a target and was achieved through the ring of black boards on the outer walls at the children's level.

Music space. Large musical notes look down on a carpet-heated area, surrounded by cast-in primative musical instruments: Glockenspiel, xylophone, bells and elastic bands stretching between 54 shiny, smooth, glass telegraph insulators. Orange and blue is the colour scheme. One blackboard to 'draw' music is cast in the wall.

Large, smooth, pottery insulators and pretty stone paving to make the verandah interesting.

Blocks and dolls, 'playing-house' space. The frieze is a skyline of familiar local buildings (e.g. the 'Barrack's Arch') together with many kinds of building forms, spires, domes, multi-storey car parks, sawtooth factory roofs and chimneys, office blocks and steel frames under construction with tall cranes.

These complemented by a great variety of little nooks, niches, shelves, slits, holes and recesses below sill level which invite incorporation into their games by the children as 'aedicules', little houses. Blackboard painted surrounds allow the children to decorate the 'house' they play with. Again, heated carpet. Colour theme is green, orange and grey.

The verandah is semi-enclosed by a row of multi-faced concrete castings with lots of hollows and other objects to stimulate and satisfy the child's vivid imagination. Tiny carrara marble paths are set in the floor.

Art area. From above window level, big free, flowing curvaceous shapes in yellow, green, blue and grey give the area its special character. A scene of wildflowers and trees contrast on one wall. At child's eye level, six large, inviting blackboards form the facets of the walls. Springy vinyl sheeting covers the floor. A pattern of rings, set in concrete, outside the verandah, act as foot scrapers. Under the verandah, in shade from the summer sun, is the sand pit, the medium for much child art.

FLOOR PLAN.

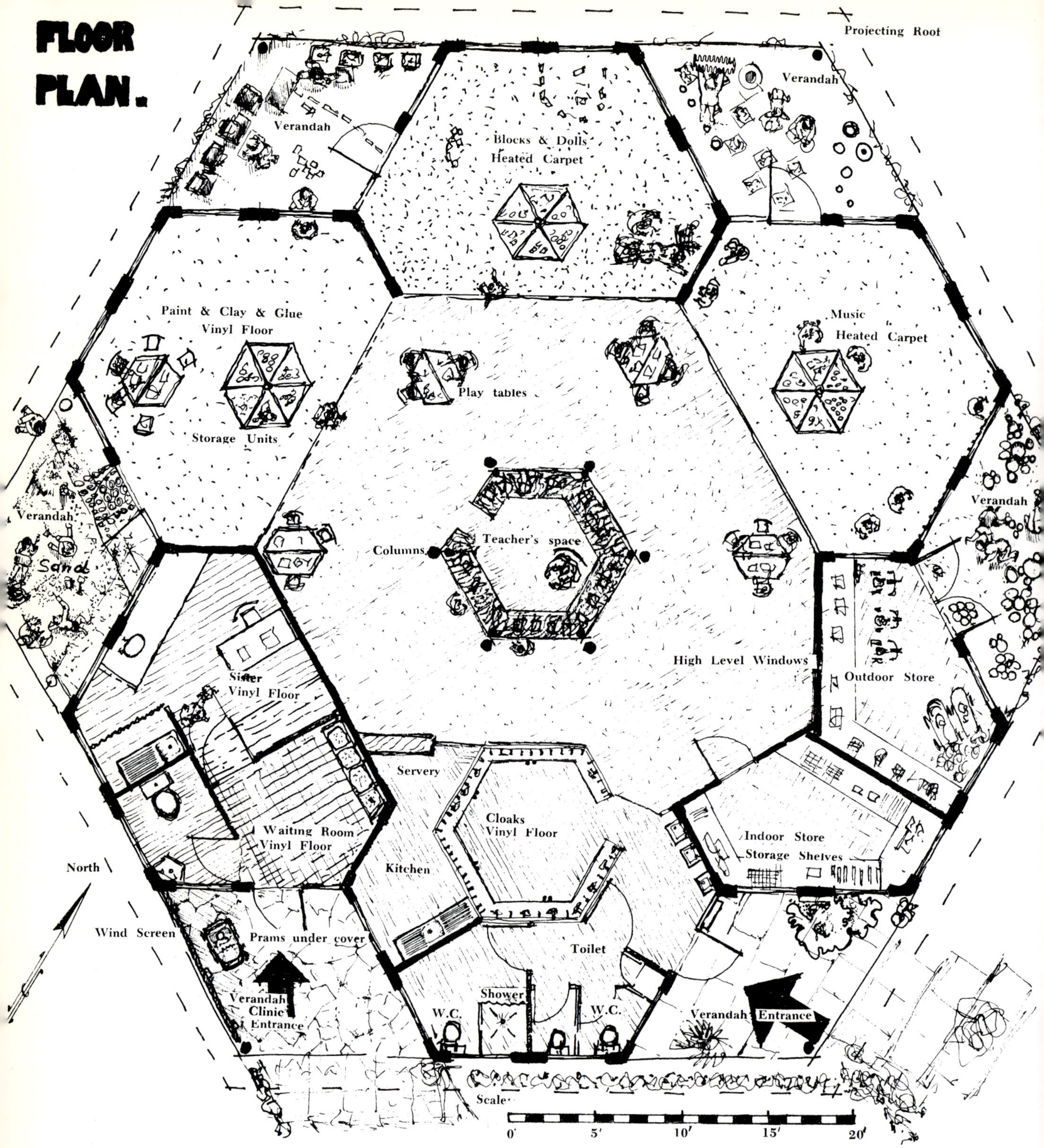

Kitchen entrance and cloaks. Sink and servery, are in a little area off the main space, with the bass-relief of a giant bowl of fruit and a multitude of kitchen utensils.

In all these 'water' spaces marine life is the theme for the bass-relief sculpture.

Structure. On a R.C. slab, thickened strategically, 18 precast concrete 'V' shaped units, weighing 1½ tons each, 17.7 m² x 12.8 m (11 ft x 8 ft) form the entire wall structure. A window opening is cast into each face. The units are joined by timber frame panels 8.8 m x 12.8 m (5 ft 6 in x 8 ft) which include the doors.

Three crucial bolts protrude from the top of each facet of the concrete units. These serve several functions:

1. To fix reinforcement for casting.
2. To hold lifting tackle.
3. To fix for transport.
4. To hold continuous wall plates 100 mm x 76 mm (4 in x 3 in).

Construction. The kindergarten was built in seven weeks from the delivery on site of the 18 concrete units. The foundation slab had been cast previously. About one-third of the actual construction of the project, by cost, was carried out directly by the

P.E.E.R. Institute. The help of parents, friends, and donations (i.e. cast iron gates, pipes, tyres, etc.) made some good things possible.

The way in which the materials and spaces are expected to perform are a clear hypothesis, against which this experiment will be tested, by creative research of the P.E.E.R. Institute.

Cost. The contract price for this combined prototype kindergarten and infant clinic, compared well with conventional building being cheaper than the usual brick building. The Sculp-crete liner art was supplied as part of the architectural design work and for those fees.

External appearance. The railway sleeper boundary walls, which avoid the usual cage like wire-mesh effect, itself offer play facilities, particularly as the zig-zag form gives not only stable structure but makes running along the top more fun. It gives a semi-enclosed look to each 'zig' and 'zag'. Very fast growing trees and shrubs will soon give shade. (Many species of tree, well watered, grow 1.5 m (5 ft) a year in Western Australia. The cast iron gates in white, blue and gold, much taller than the fence are a symbolic welcome. They face onto the access road, which is to have a bay for bus stop and some parking and onto the path.

Chains acting as 'down-pipes' create temporary waterfalls.

Feedback. Neither the kindergarten authorities nor the local authority took the creative experimental attitude of watching the use and encouraging the most advantageous use of what was provided. The local authority removed the magnificent and well-loved fence on quite irrational and hateful grounds, and replaced it with a wire fence. Just to keep people out this could have been placed on the outside of the sleeper play wall. Waste of a costly and valuable asset. The magnificent cast-iron gates were removed. All this was done without reference to the architect who had obtained these, free of charge, transported them, and presented them to the kindergarten.

The interior friezes invited the parents to get out their brushes and paints and decorate them in a style that nearly gave the architect a fit when visiting the building to show it off to a classy Swedish visitor. However, it invited participation.

The musical instruments built into the wall were never repaired — a simple task built into the design — but had decayed to quite some extent. The remnants were still things of interest and evocative to the children's imagination.

Beyond that it could be seen that a teacher can overlook all sorts of advantages in a new design if, as we tend to train our students, they follow blindly past stereo-typed behaviour. On the other hand the alive teacher can make and will make more joy and

Early, crude Sculp-crete techniques operated by Ralph Hibble (top), Erica Ritter and Paul Ritter, on the V-shaped units in the casting yard. The various types of friezes can be identified.

use the opportunities allowed for and encouraged by the design.

The sense of belonging of the parents, in the first instance, was to some considerable extent due to the individual child-orientated nature of the building. A good effect for the whole neighbourhood.

Participation. Studying the many uses of any room both aesthetic and functional in the more normal sense, there are countless features that can be incorporated in the concrete casting by forethought in formliner treatment. Our Rockingham Park Kindergarten is a good example of this. This also ties up with the study I made of the bio-energetic implications of Summerson's aedicule theory, the simple fact that recesses in walls have always been attractive **per se,** not only because they are useful, for this or that display or storage. We used this approach with good effect and excellent economy at Rockingham.

VI.5.3. Childrens' Protection Centre.

Four panels (overleaf) for an existing facade of a Kindergarten in Perth. It was found too stern by its management which required that it be made attractive and alive for children and passers-by. Carried out by Paul Ritter, Ralph Hibble, consultant in technology and Simon Miller, consultant sculptor.

'We love you' frieze of happy children's faces right around the Kindergarten at Rockingham, W.A. Sculpted by Mark White after an idea by Paul Ritter. The three illustrations show the effect of different light on what is a green glazed surface and the crude colour of the formwork lining prior to casting of a snake. Theme mould held by Jean Ritter, Director P.E.E.R.

Scale:

←— 20 cm —→

To individual taste chandeliers, flower pots, waste paper receptacles, vases, lamp bases, can be made to matching themes. Thus a hotel rooms' ancillaries, for example, can match in style and pattern, plain or fancy.

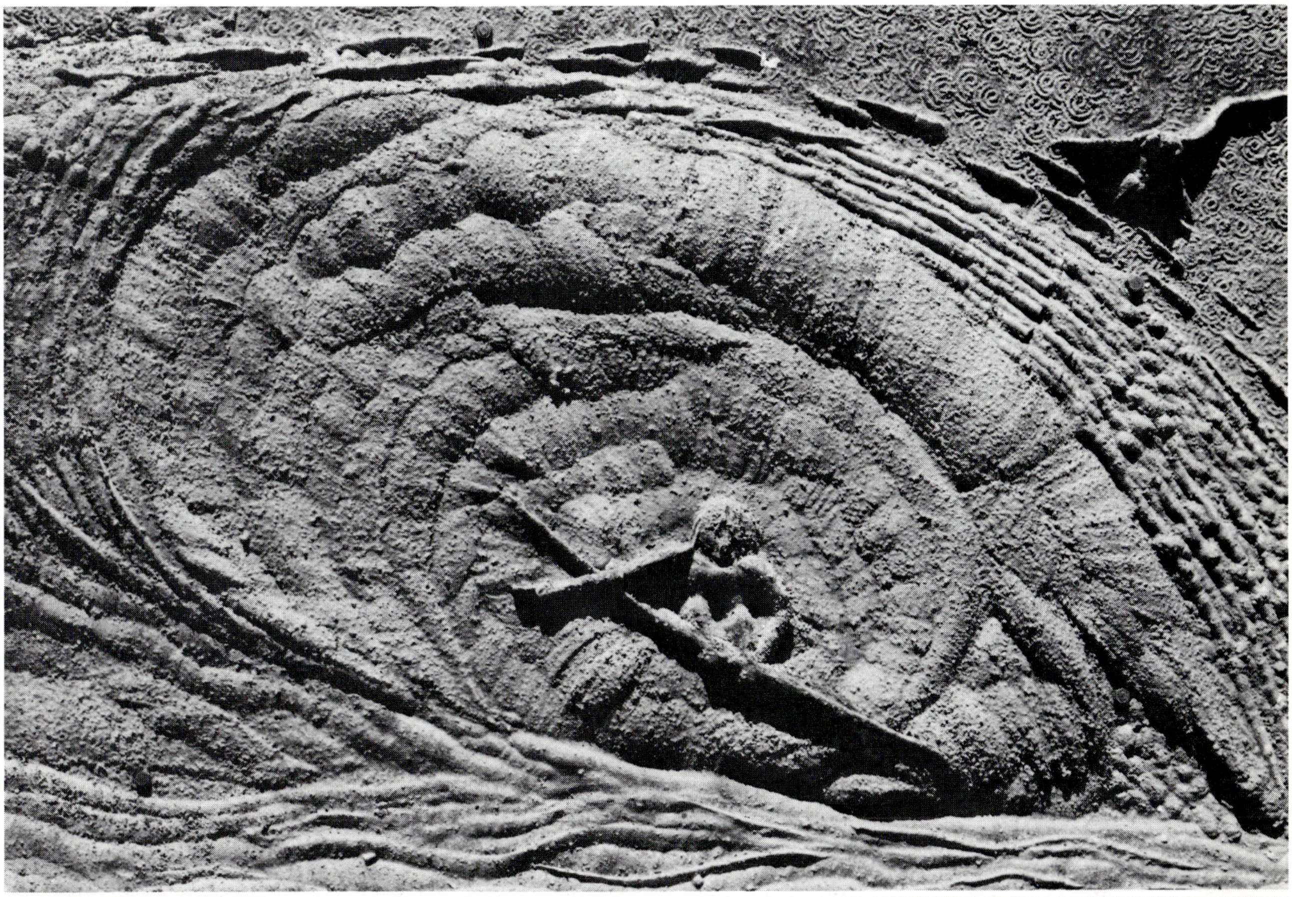

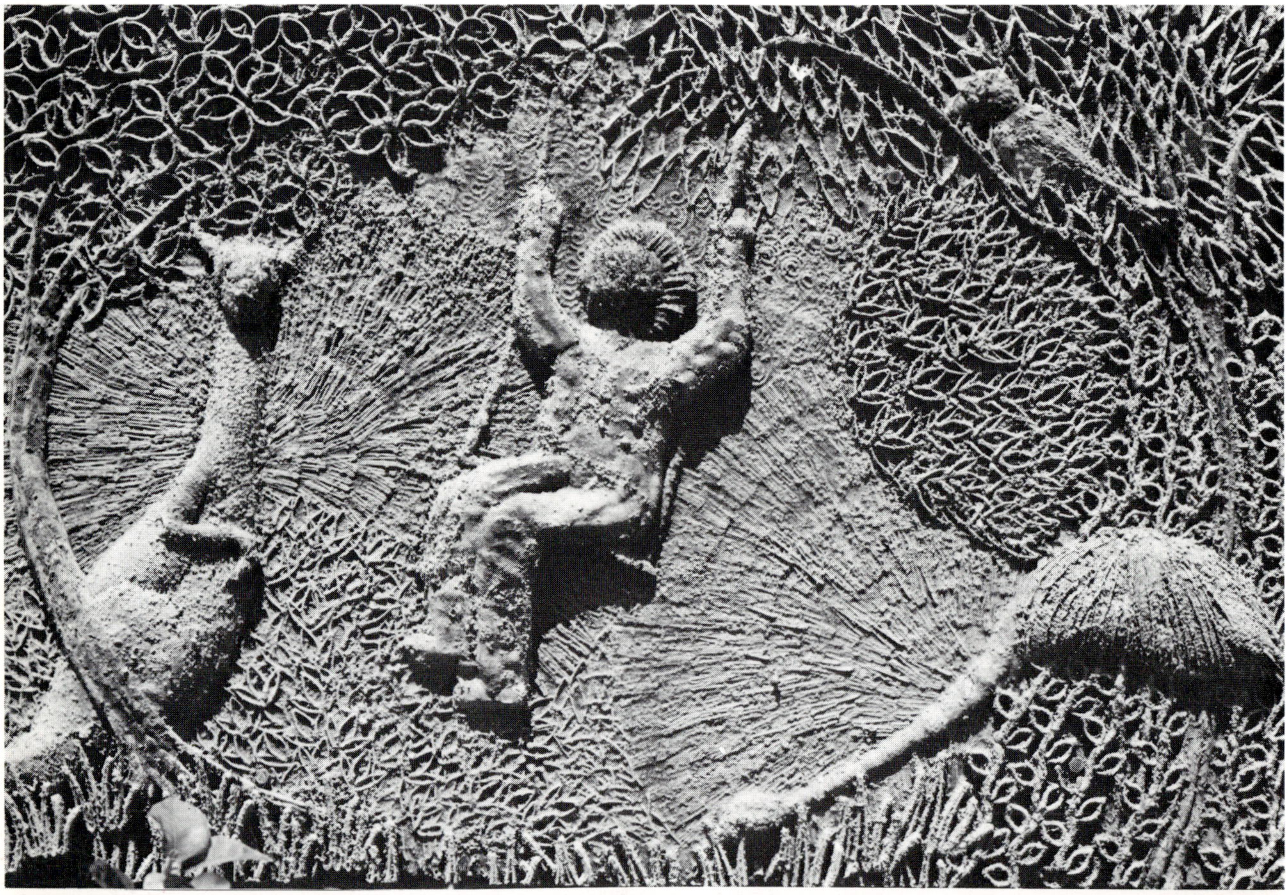

VI.5.4. National and regional character in architecture

In Iran, as in many countries with rich cultural histories, there is an understandable desire to draw on this heritage for inspiration in architecture. The world wide change from hand work, because of its expense and slow nature, make it impractical to emulate the symbols, patterns, surfaces, shapes, textures.

Because the moulds for Sculp-crete are so easily and quickly prepared, as are the tools that make them, and their combination for mass-production, this invention opens up a new world of possibilities, of drawing on the old world for inspiration.

It was sad that the change of Government in Iran, which was essentially a change to the traditional, postponed indefinitely the exhibition by P.E.E.R. organised by the Iranian Government to demonstrate with Iranian school children how Sculp-crete makes it possible for them to use Persian & Islamic sources for their creative work.

The economy and basic simplicity of the techniques involved and the combination of the artist's creativity and bio-chemical processes, ensure a richness and variety with outstanding speed and ease, unknown to previous cultures.

Traditional patterns, textures, colours and glazes can be translated into the Sculp-crete medium. Designs of the past gain a new vitality and personality. Concrete is humanized.

The examples were produced by a team of P.E.E.R.: Paul Ritter, Director, and consultants Ralph Hibble and Simon Miller, on the basis of research, to indicate the range of possibilities.

The panels 80 x 120 cm cast in fibreglass reinforced concrete (Cem-Fil) are one to two cm thick and weigh on average 20 kg.

Alternatively the Sculp-crete method allows such designs, or over-all patterns, to be cast in place, during the normal pouring of ordinary structural concrete, the formliner protecting it, until the completion of the job.

Prepared polystyrene mould for bas-relief after Codex Mixtec shown cast. Note richer colour of mould.

After lacquered box produced at Olinala, Mexico.

Coloured polystyrene mould for above panel, ready for casting and with its own charm.

After Islamic pattern.

After Fabric from Valle de Chaucay, Peru.

After Islamic pattern.

After Tiki Window Design, New Zealand.

Coloured polystyrene mould for above.

After Board wood face mask, New Zealand, Oceania.

After Codex of the Mixtec culture Mexico.

After Temple of Karmok, Egypt.

After rock bas-relief, Tomb of Ti at Saggara, Egypt.

After Indian Cloak, Mountain Goat Hair, Tlingit Culture, North America.

After plush-like mats woven out of rafia palm leaves, Bakuba Africa.

After cloth design for the Sumba family, Indonesia.

Scale:
10 cm

Indian cotton pattern. The tools with which the printing is done have a simlarity to Sculp-crete tool elements which allows the transposition from cloth to concrete and from flat to basreflied quite simply.

Scale:

10 cm

One of twenty-four panels making up a garden wall. Interspersed with plain bio-wall panels. The 2m x 1m panels were cast in sections, easy to slip into u-shaped vertical supports. The molds, to ensure continuity of design, were prepared as one piece and then cut for casting. At 70 Regency Drive, Crestwood, Perth, (a famous suburb bio-functionally planned by Paul Ritter), W. Australia.

Scale:

10 cm

Patterns of considerable depth can be created by sticking the patterns cut out of the mould onto other areas of the mould. In this case the theme is hexagons.

Carving in three dimensions with the movement of the whole arm, represents a new technique in art, with countless opportunities, therapeutic educational etc.

Scale:
10 cm

Superimpositions of traditional square spirals gives a rich and satisfying classically based pattern.

Self-supporting garden wall unit not requiring foundations are a Sculp-crete product with decorative textures or patterns on both sides.

The 'Cross of tears and hope', a sculpture presented to the people of Darwin after the cyclone disaster of 1974. Now in the Town Hall. The amber glaze catches light in a dramatic way.

Opera singer Rita Streich chose the sculp-crete for her home. She saw it as 'pure music'. Sculpture by Ralph Hibble.

'Space chess' a sculpture by Paul Ritter. Combination of fine white concrete with lamp-filaments and the Sculp-crete glaze leads to a luminescence and refinement unequalled in other modern techniques. The illustration is approx. two-thirds life size. The figures 'float' on a hyperbolic paraboloid surface.

'Curves of Life' — a 27.3 m (90 ft) long bas-relief by Paul Ritter, produced with one assistant in two days for an existing office block in the main street of the capital of Western Australia, Perth. Concreting was done with normal mix. Erection of the nine pre-cast panels was done in half a day. The Sculp-crete technique showed a staggering economy over any other similar bas-relief effect.

The Outside, the inside, and the bottom of this urn are decorated with re-entrant patterns. A tour-de-force in concrete.

VI.6. Examples designed to satisfy the visual, tactile, atmospheric, life-supporting, social, and creative participation needs.

VI.6.1. Architect Participation in design.

The masons of the Middle Ages sat all round the cathedrals carving 'their thing'. Judging by the incredible variety of carvings, sacred and profane, that adorn cathedrals, it is a pity that many are too high for the naked eye. That may explain the freedom of taste.

Architects have longed for more direct participation, harking back to the unity possible and managed during the Renaissance. Then, magnificent works showed the artist able to conceive his built form and its decorative content homogeneously. Not one stuck to the other afterwards, as has happened in the vast majority of cases in recent times. The early dearth of detail and decor in modern architecture, has lead to profusions of stuck-on things in every shape and form and material and quality and size.

Le Corbusier's tiny shells were a lovely touch. It awaits recognition by the profession that this was the finger, slender maybe, pointing to a new integral style. A style using the plastic nature of an astonishingly cheap and strong mixture — concrete a for runner to the Sculp-crete philosophy.

Styles spring from a collection of factors out of a forming culture pattern incorporating the state of technology, values, symbols, customs, politics or government, philosophies. Unlike fashions they represent something substantial and have their roots deeply anchored. In the short history of man a small number of styles has suggested that out of this is born a recognizable matrix of choices related to the current criteria. I believe that the requirements and awareness crystallizing out of the twentieth century into the twenty-first, are forming criteria reflecting the bio-technic culture patterns for the third millenium.

As the participation of masons complemented the grandiose structural concepts of the cathedrals, so the desire for participation in our bio-functional future is already showing itself in the writing and even work of people round the world. Exceptional and eccentric as their work may seem it is in participation of many kinds that the character of the bio-functional style will develop: an example is Leverents' great architecture, in Sweden: loved into being, inch by inch, with the architect on site, even as a very old man, coaxing the workman to be his extended arm. Like the best architecture — can't be appreciated unless it's experienced, not just photographed. He did not care for publicity. His work is rarely illustrated. Not part of the architectural fashion world.

The bio-functional needs listed and described will be absorbed into the architect's vocabulary so that his artistic expression will form a complement and express and combine in new and delightful ways man's creative abilities. The ease of doing this with Sculp-crete, the infinite choice to adapt pattern, texture, and form to needs, gives the chance. The architect in the factory or on the site can love (or hate) his formwork into shape. The architecture resulting will be integral: indivisible, functionally and structurally a part of the building. Not stuck on, but grown, as the bark of a tree develops, with it.

Just what will all this look like is quite beyond me to even postulate. The wonder of being alive would be reduced if I was a clairvoyant and able to see it all. Unlike the more 'formal', classical, limited styles of the past the riotous excesses, the diversity, the choices will be immensely greater reflecting our technological brilliance, our immaturity and the much-increased variety of all phenomena in the current and emerging world.

How the elements of the style may emerge has been indicated by taking one function at a time and adding a further component. For sure it will be a functional style, it will show to have some of its roots in the Bauhaus but grows beyond just because their realization of functions was limited. Paul Klee's expressions of think-feel-fun and life was not reflected in the walls, it was stuck on to the 'smooth functional walls' that collected no dirt. Now this sort of thing can become part of the fabric.

Carving by the movement of the arm and wrist in formwork can make it possible for base-relief designs to be created by the architect himself, for an enormous building in one or two weeks' work. Immense, fresh degrees of scope, geared to our technologically brilliant era as well as our desire to think and feel on a big scale 'Eco-architecture', 'traffic architecture', ekistics, three words recently coined to indicate that not only **can** we now build enormously large complexes in the centre of our cities, like the transplanting of organs or limbs, but, we can think it through in appropriately penetrating and effective ways: Creative throught not stifled, but inspired by building material.

VI.6.2. Client participation.

To get people involved in the design or patterning of the surfaces, at school, home, and in community, to get the architect in a position where he can use the formliners or moulds they produce these are real practical ways open to humanity to get some of the involvement and variety for which so many are longing, judging by much comment all round.

Some indications of how this is possible and what the range might be, arise out of our experiments with young children at a primary school, the general public at a creative participation exhibition, and fourteen schools in a commemorative project.

VI.6.3. Sculp-crete for schools with primary school children average age 8 years.

From a small state school, Canning Shire, in the Metropolitan Region of Perth, Western Australia.

(i) The veranda wall to be created for weather protection, along the existing columns, was of a total length of about 30 m (98 ft) and the concrete panels to form it were to be decorated on both sides. Vertical costing in a bank of 36 panels in one operation.

(ii) Hypothesis tested by experiment
That it is practicable to co-ordinate architects, administration, school staff, parents, children, recording media, with factory or workshop to let children design and produce the moulds for concrete walls for a school, the whole installation to cost no more than traditional brick walling.

(iii) Outline of experimental procedure

1. Introduce Public Works Department to method and potential of Sculp-crete.
2. The same for the Education Department from their point of view.
3. PWD inquiry for cost and feasibility of Sculp-crete for verandah wall.
4. Submit costs and make agreements.
5. Headmaster co-ordinates teachers, time-tables, parents' notes, transport.
6. P.E.E.R. Institute organises one day work sessions for children at its workshop with colour television film record as news item.
7. Manufacture of panels in moulds as sculpted by children.
8. Erection.
9. Communication and Recording.
10. Official Approval by Minister of Works and Deputy Premier.

Electrical, safe low voltage, instant heat imprinting is the main technique for shaping molds in expanded polystyrene as required.

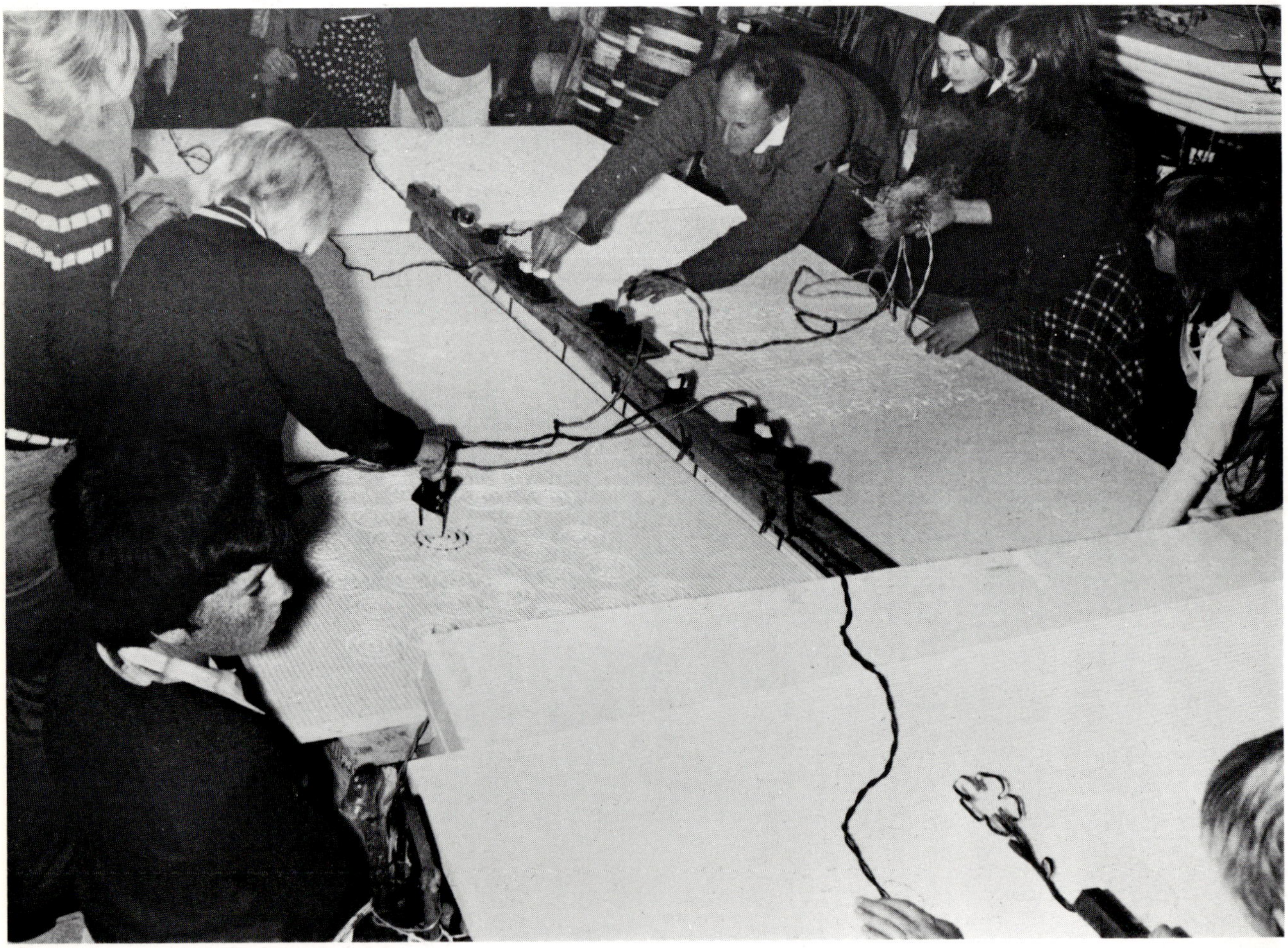

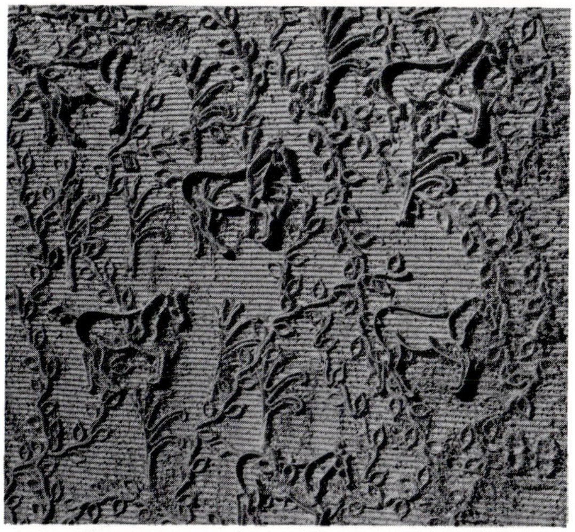

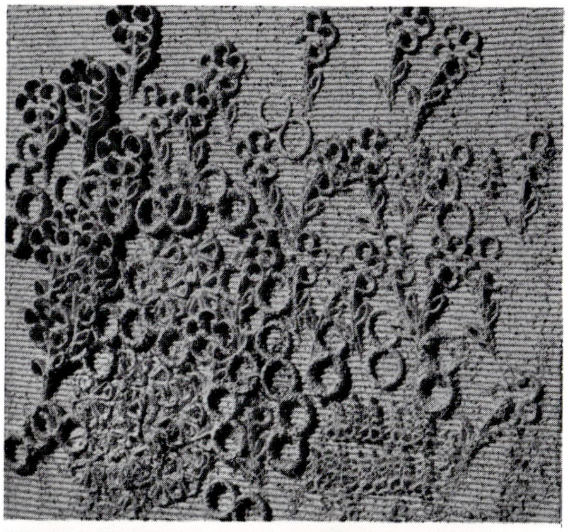
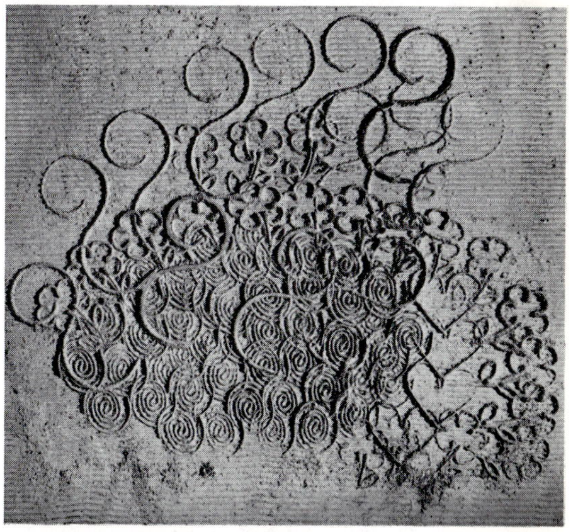

(iv) The key stage of the experiment
On the appointed day the children arrived at the P.E.E.R. workshop in four parties of sixteen, in age groups, two morning, two afternoom.

They were introduced to the simple process and almost immediately started to imprint choosing, from some 50 tools. This attractive activity captivated very quickly and there was no shortage of ideas, individuality or joy as 72 liners were produced in the one day.

The youngest groups got into their stride quickly, older children and teachers, showing hesitancy It is important to note that all children who had parents' permission came, they were in no way selected for artistic merit, talent, etc. The children chose what they wanted to do. Interest was focused on the job in hand so that there were no problems of confusion or what is called 'discipline'.

The wonder at themselves shaping their own school became real when, unceremoniously, the concrete panels were placed in position while the children were at school, coming out at lession breaks to crowd around, with 'there's mine', 'looks, there's yours', as they recognized the positive cast of their negative initials on their work for posterity to see.

Parents became specially involved through seeing their children working at the workshop on the television news.

It should be noted that Sculp-crete is, in its concept, a non-perfectionist technique. Although this has been widely recognized as the only appropriate aesthetic approach to concrete in general, after decades of frustrated perfectionists, it is important that this characteristic helped enormously in allowing the children freedom and joy in their efforts, to make the exercise feasible.

The boldness of the aesthetics of Sculp-crete ensures a remarkable unity from the many diverse panels.

(v) Conclusions
The experiment gave clear evidence that the hypothesis was correct:

1. Co-ordination of many bodies was possible.
2. The children produced good creative work, school and community merging in the educational process.
3. The work was done at the cost budgeted and it became clear that in future substantial economies on traditional building costs can be made.
4. The relationship of child to his school environment is improved.
5. It was clearly seen that many varied kinds of such participation are feasible that the method is open ended and invites imaginative exploitation.
6. The commercial communication media, the official attitude to the finished product, the response to films shown in Western Australia and overseas indicates a general apprecation and support for the principle of people and children having a hand in shaping their environment.

VI.6.4. Sculp-crete school kit.

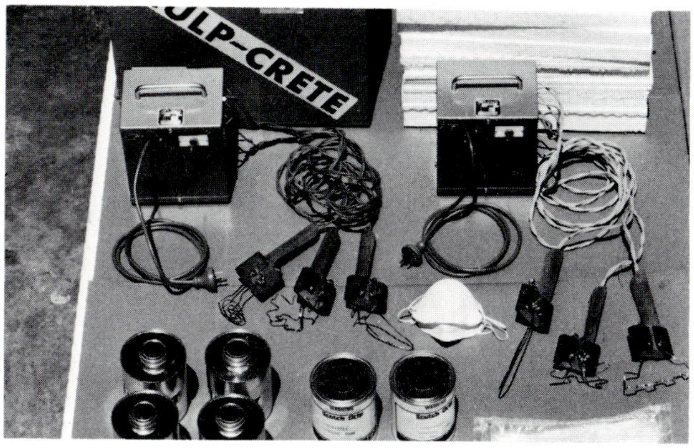

The criteria for the design of the Sculp-crete School Kit were safety, particularly against electric shock; robustness & simplicity to allow easy repair of wilful damage; opportunity for easy creation of individual tool elements; flexibility and practicability for class teaching; easy replacement of back up supplies with economy.

The immense enthusiasm and success of the work with schools led the P.E.E.R. Institute to develop a Sculp-crete School Kit in 1979. The intention is to allow children of all ages to enjoy the new art and craft form potential without leaving their schools, as part of the normal class teaching and as part of the curriculum where this is relevant. The top art advisors of the major states of Australia supported this move and the adoption of this highly satisfying outlet for creative effort. The technology is now spreading to other countries with the knowledge of Sculp-crete and the popularity of participation.

As children become familiar with the potential of concrete and realise the ease with which beauty can be created as part of the new bio-functional architectural style, this will act as a catalyts to architects. Many people will emerge from the schools, both lay and professional, no longer afraid to participate with feeling in the unification of structure and decor so that more whole and wholesome buildings emerge. Private dwellings, urban spaces, communal buildings, and commercial buildings will be affected.

A typical primary school application of SCULP-CRETE in 2 lessons.

Costume Jewellery

Attach, say, the spiral element to tool handle. To control heat — press switch and let the heated element indent its spiral shape into the thin polystyrene sheet by about 3 mm. A row of spirals takes a few seconds. Mix up a little tint with the 3M glue. Roughly paint into grooves of the spiral form. Mix a tiny bit of cement with twice as much sand and a spoon or two of 1 cm long fibreglass cut from the length provided, then add water gradually until the mixture is just right to squash and prod into the grooves. Let the spoon bounce up and down on the polystyrene to vibrate the mortar mix into the tiny mould spaces to get coral effects. Stick the ends of a fuse wire loop into one of the mortar filled grooves. The loop will serve to thread the jewellery. Place into plastic bag for good curing till week later.

Second Period: lay polystyrene on wire cake tray, spiral mortar shapes at the bottom, with some newspaper or a flat tray under to catch very sticky extra glaze. Sprinkle Sculp-crete solvent gently through the perforated screw top with plastic inlay. Instantly the polystyrene melts and the colour glazed jewellery appears shining on the tray. Leave to dry making sure that the sticky glaze does not bond the jewellery to the tray by lifting a few times, very gently, with a pencil as it dries rapidly. More glaze, — a mixture of polystyrene, colour, solvent, — can be ladled selectively for richer effects.

As elements are easily shaped each person can have what he would like.

Containers, costume jewellery, table ornaments, wall plaques, lamp bases, picture frames, ashtrays, house numbers, name plates, crests, paving slabs, tiles, free standing sculptures, jig saws, Christmas decorations, vases, flower pots, fruit bowls, trays, book ends, post boxes, paper weights, doorstops, dog dish, etc.

Suggestions for Sculp-crete work in schools

Explain the idea and ways of Sculp-crete simply. Give practice with various tools and scrap polystyrene. Get the feel of the various tool elements for the shapes and patterns you are trying to achieve. Then let the tool work for you by noting shapes cut out.

Learn to think and feel 'back to front', i.e. in the negative, for mould preparations. A useful way is to imagine you are sticking out your tongue and similarly whatever you press in or carve in will stick out, towards 'them'. Once the imagination has made that flip it is relatively easy to feel and create the forms in the negative without having to be thinking of this all the time, cramping the creative flow. Quick-setting plaster casts can help to show students what is involved.

Conceiving what is wanted before working polystyrene should be, I believe, a feeling experience — whether its an ash tray or a human figure, whether ridiculous or sublime. There is a distinct limit on how much can be sketched on a flat surface for carving, particularly when the aim is smooth flowing movements of the wrist and the arm.

Public participation through spontaneous use of sculp-crete tools, for a small charge, by the public at a participation exhibition. The moulds were used to build a hexagonal children's house from v-shaped units. Contrast of style and talent are fused into a remarkable unity by the sculp-crete look.

VI.6.5 'Hex'

So keen is the desire by people to participate that they paid half a dollar to be able to use our tools for a little while to create their imprint on the moulds that were to be used for a children's house, and cast in Sculp-crete. From consultants to kids, artists professional and amateur to folks who never draw or sculpt, the method enticed them all to participate on the same surface. The strength of the character of the Sculp-crete technique has unified, to a remarkable extent, the very varied shapes, forms, styles or and qualities of expression.

VI.6.6. Perth Pageant

Perth Pageant is an educreational application of the Sculp-crete technology and art, by the young of Western Australia, to celebrate the 150th birthday of the State.

It is a permanent contribution to the townscape of Perth co-ordinated by the P.E.E.R. Institute, bringing together the creative talent of fourteen schools, State Government and private enterprise.

The eighteen moulds were produced, (after the preliminary research) on average, in one and a half days, by teams of about four who had only one short preview of the whole process. Concreting took a day. The metamorphosis of dissolving the mould, and so transferring the colour and glaze onto the concrete by gravity, took a mere hour, for seventeen 2 m^2 panels.

In tribute to the fourteen schools it must be said that they co-operated fully with only 24 hours notice on this project and in response to just one enthusiastic phone call to the Principal, by Paul Ritter.

The enthusiasm caught on and reflects in the success of the project, as teachers and artists turned up for an explanation and demonstration, making up their minds there and then, they wanted to 'be in it'. spontaneity was the keynote of the pleasure taken by all concerned and some worked on weekends to get their panel mould completed.

We co-ordinated the efforts of small country towns and capital city, of primary schools to final year university students, self-regulated community school to highly disciplined state and private colleges; the project showed that the idea is widely applicable. As varying groups worked side by side in close contact, the meeting of various differing groups was of interest and value.

In the main the students arrived with rough sketches and montages of the detail various students had independently studied. It had been explained that the compositions should have the boldness which would give impact and meaning from 30 m and yet have the detail to delight the eye from close by.

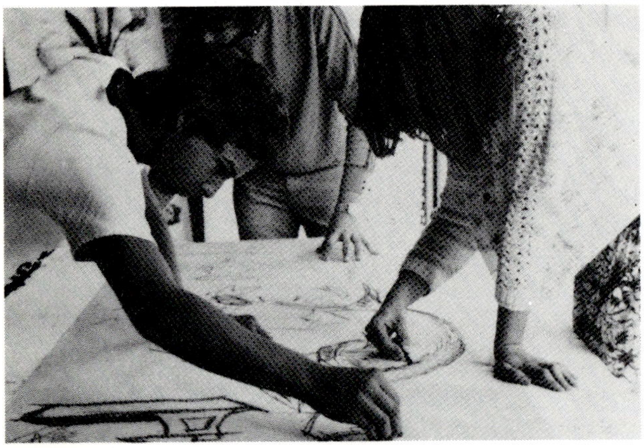

Having chosen a subject, the students study its history and the shapes and forms involved and prepare sketch compositions to fill the panel of approximately 1 m x 2 m. They know that the coloured and glazed bas-reliefs will be viewed from a distance as well as close by.

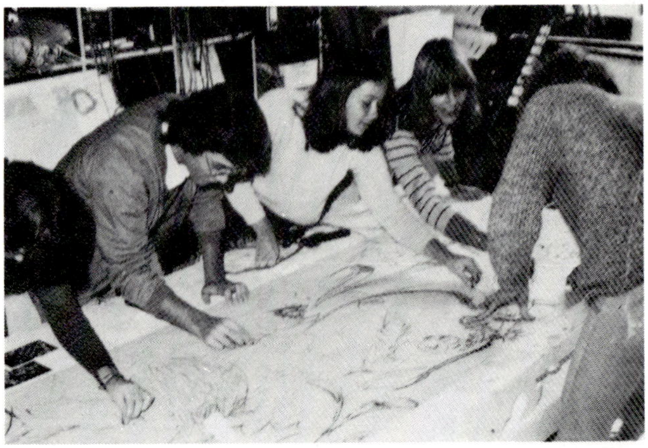

The sketch cartoon is transcribed to the polystyrene block out of which the bas-relief mould is to be shaped. Only basic outlines are useful, as the majority of the surface disappears when shaping proceeds.

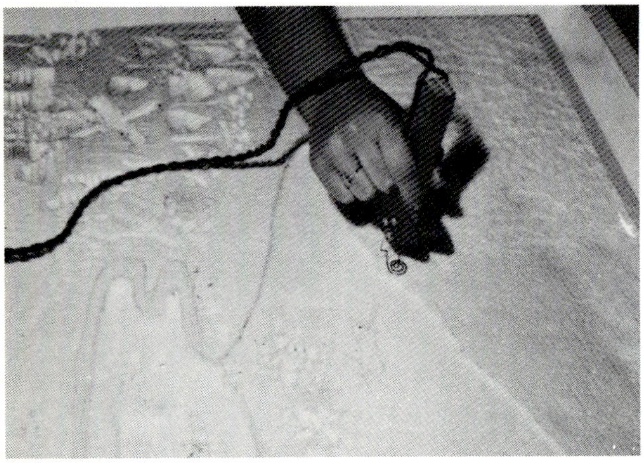

The students learn to think inside out, as they shape the polystyrene with low voltage electric heat tools of many shapes. The bas-relief concrete will fill the form of the mould. Every hole becomes a bump. Undercutting is feasible and easy as the concrete penetrates through vibration, during casting, into all cavities.

The painting by brush and spray gun need not be accurate. The first layer of paint is the only one that shows finally. Where the grey or white of the natural cement colour is required, no colour is applied. The moulds, because of the white polystyrene and bright colours, look bold and garish.

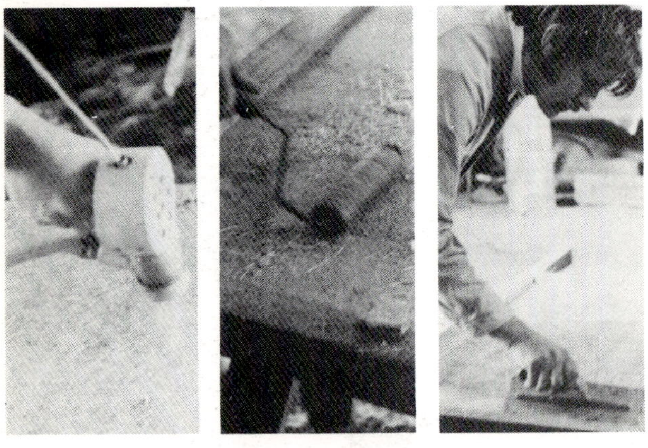

The panels are sprayed and cast in ultra strong fibreglass reinforced concrete for thinness, lightness.They are vibrated, rolled and tamped. Hanging devices are included. The work was done with MUNILOC at their factory. The concrete was then allowed to cure for a week and to dry out for another three days in the sunshine.

The metamorphosis, from white expanded polystyrene mould to colour glazes, through spraying the surface with solvent, was a brilliant spectacle, attended by representatives from University. Parliament, local authority, the schools, the artists and their parents, and televised by many cameras.

Students painting the white, heat carved and shaped polystyrene mould. This first stage of the Sculp-crete process can be the finished result is fo intended: An extremely light but limited application.

The extremely light polystyrene, heat carved and coloured moulds ready for casting: In themselves they were a most impressive, large-scale, brilliantly coloured adornment produced at the rate of one a day by the students of various schools.

Typical detail of sculp-crete texture

ABSTRACT HISTORY by University of W.A., School of Architecture, 5th year, Perth. Students: Michael Rokich, Nick Broadhurst, Dov Zivkovic, Peter Beyer. Limiting themselves to the natural grey concrete colour there lies the superimposition of natural elements conquered increasingly by the larger scale artefacts of man.

Differences of colour through lighting is emphasized by the glazes.

A heroic deed by engineer C. Y. O'Connor linking the parched Goldfields with a water pipe-life line to the overflowing dams of the richer rainfall areas of the coast. Paddy Hannan first found gold at Kalgoorlie. The statue commemorating him can be seen on the left side of this half full size detail of the 3m x 1m horizontal panel. This panel was the proto-type, done in solid reinforced concrete. It weighs six times as much as the other panels. The advisory group of the 150th Anniversary Celebrations Executive Committee, State Government appointed, were unanimous in their views: 'This panel is an outstanding good example of this art form and that, if such standard could be sustained in a series of panels, then Perth and Western Australia would acquire an asset which was artistically unique and pertinent to the 150th anniversary of the State and the City of Perth'. The promise has been fulfilled.

THE PERTH—KALGOORLIE 300 MILE WATER PIPE LINE (1898) by Cannington State High School, Perth (Detail). Students: Winsome Hodgson, Sharon Fitzpatrick, Giselle Downs, Mark Lambert, Mark Stahl, John Rapattuni, Max Schrimps, Michael Szczyglak, Hans Mulober. Teacher: Keith Rendell.

HISTORY OF NEDLANDS —
DETAIL by Seniors' Presbyterian Ladies College, Perth
Students: Claire Paton, Nicola Patrick. Teacher: Mrs J. Mell.

The artificial coral shown below is a particularly rich example. It is produced through the penetration, by vibration, of the fine cement slurry from the concrete, into the narrow space, which is left around each globule of polystyrene. The multitude of these, holding together, make up the solid polystyrene slab. Just as natural coral is an external skeleton, grown around a globule of life, so the artificial coral is similarly formed.

THE GOLDRUSH, KALGOORLIE (DETAIL) by Eastern Goldfields State High School, Kalgoorlie. Students: Anne Ribarich, Cassie Bussell, Jennylyn Quick, Suzanne Bain, Vanessa Norris. Teachers: Mr Peter Clarke, Miss Robyn Dwyer. The characteristic shapes of the early implements and early miners are brought into a rich perspective.

WEST AUSTRALIAN WILDFLOWERS (DETAIL) by Balcatta High School, Perth. Students: Robyn Cant, Lloyd Kan, Mathew Darcey, Leslee Rowlands, Karen Lawrence. Teacher: Mrs Paula Summerville. A breathtakingly bold tour de force, of some of the astonishingly varied and picturesque botanical riches of the State. The sculp-crete technique is used to the full with appropriate and fine effect.

PROTESTING CROWD (DETAIL) by Community School, Fremantle Students: Sam Lemann, Ric Bell, Tom Game, Gail Evans, Annie Davy, Jo Dudden. Teacher: Jane Liddon

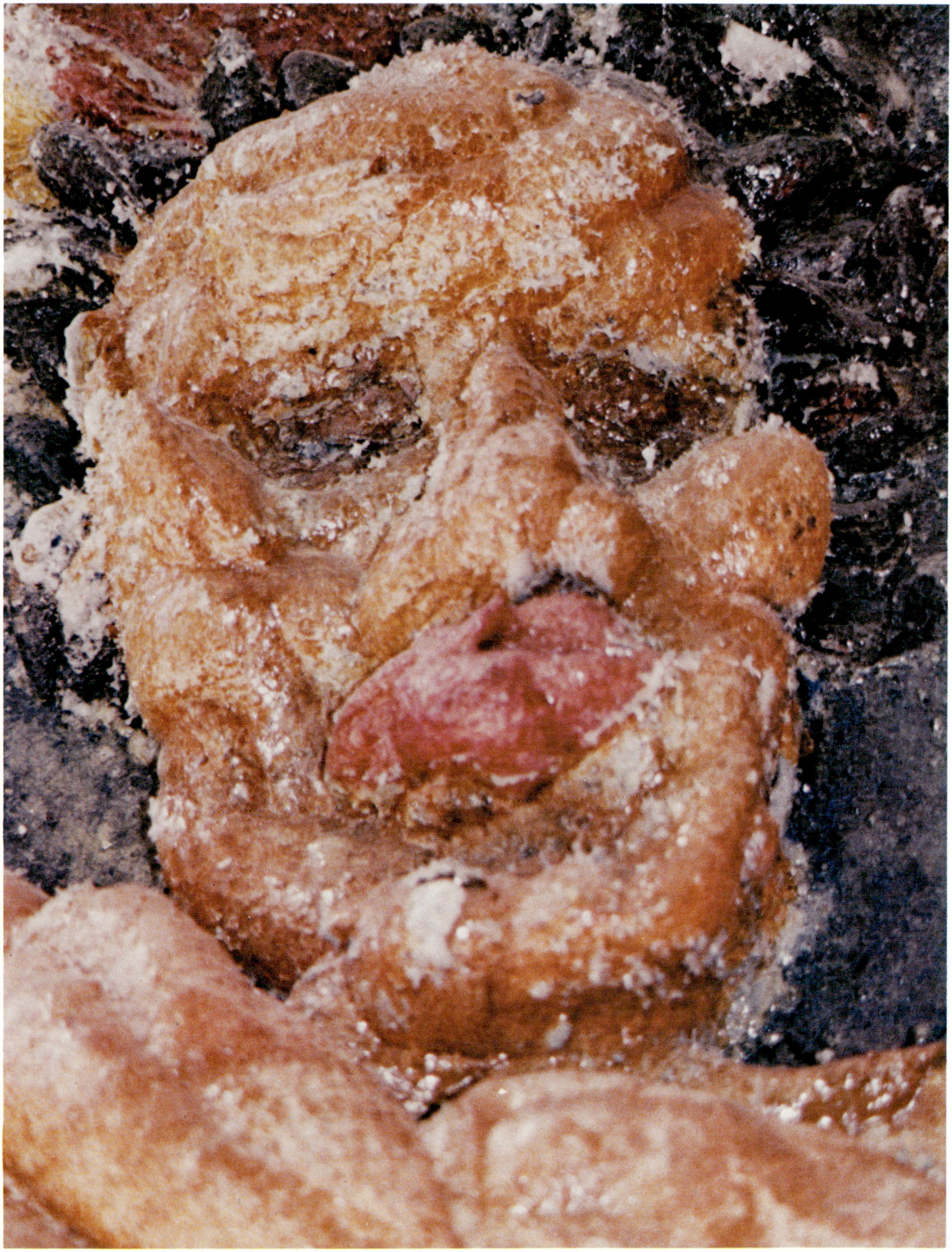

SWANBOURNE BEACH by Community School, Fremantle.
Students: Sam Lemann, Ric Bell, Tom Game, Gail Evans,
Annie Davy, Jo Dudden. Teacher: Jane Liddon.

FREMANTLE PORT by Applecross State High School, Perth Students: Tim Keevil, Megan Panfull, William Wood. Teacher: Mrs Rosemary Wallace.

OIL AND IRON by Hale School, Perth. Students: Glen Shreeve, Peter Hanlon, Richard Blackman, Simon Brownfield, Jeffrey Dye, Evans McDonald. Teacher: Andres J.C. Foss.

THE FISHING INDUSTRY by Balcatta State High School, Perth. Students: Robert D'Arcy, Joanne Turnbull, Jim Watson. Teacher: Mr Phillip Stanley.

SYMBOLIC MAP OF NORTHAM, YORK, TOODYAY, COUNTRY TOWNS DISTRICT ALONG AVON RIVER by Avondale Primary School, Northam. Students: Denise Walker, Athena Marinis, Kim Clothier, Natalie Veal, Deal Every, Shane Cohen, Simon Northey, Darryl Stephen. Teacher: Les Hammill. The naive and picturesque approach catches the charm associated with early map making on the one hand and primitive Chagall like work on the other. Bridge with boats under, shows the spectial delights of bas relief. The textures, particularly along the river's edge, are remarkable.

Details showing childish delights and adult results.

THE STORY OF PERTH IN ITS BUILDINGS by Thornlie State High School, Perth. Students: Kim Ward, Judy Maras, Sarah Wollaston, Julie Cahill, Pamelia Jeffrey, Carol Frein. Teacher: Miss Jasek. The detail in this panel is derived from careful study of the history of buildings in the short 150 year history of Perth. Its delicate textures invited that it should be case in white cement. There is a sense of fun in the cars careering over and under the freeway bridge.

Marble entrance hall of 44 storey building in central Perth.

ЅWANBOURNE BEACH by Community School, Fremantle. Students: Sam Lemann, Ric Bell, Tom Game, Gail Evans, Annie Davy, Jo Dudden. Teacher: Jane Liddon.

RAPE OF THE FORRESTS by Scotch College, Perth. Students: Michael Carlin, Jamie Artelaries, Roy Gray, Neil Bailey. Teacher: Mr Brian Latham.

VLAMING DISCOVERING by Guildford Grammer School. Students: Mark Hiller, Graeme Hall, David Stacey. Teacher: Mr Barry Gardner.

HISTORY OF RAILWAYS by Bruce Rock State Junior High. Students: Leonie Vaughan, Ian Baxter, Karen Fuchsbichler, Janet McCall, Suellen Ponfrett, Michael Fuchsbichler, David Hart. Teacher: Mr C.D. du Boulay.

VINE by Juniors' Presbyterian Ladies College, Perth. Students: Beth Mell, Jenny-nn, McLellan, Nicole Lobry e Bruyn, Liz Adamson. eacher: Mrs J. Mell.

THE PARMELIA'S LANDING by Trinity College, Perth. Students: Paul Darvin, Jeremy Pownnall, Shaun Alvar, Craig D'Crue, Michael O'Connell, Paul Keenan, David Rigoll. Teacher: Adelphe Smith

PROTESTING CROWD (pictured on its side) by Community School, Fremantle. Students: Sam Lemann, Ric Bell, Tom Game, Gail Evans, Annie Davy, Jo Dudden. Teacher: Jane Liddon.

PERTH-KALGOORLIE WATER PIPELINE (pictured on its side). See pg. 2 for artists.

The coloured polystyrene moulds ready for casting the Perth Pageant panels, highly decorative in their own right. Size 90 cm x 270 cm. Each one prepared on average in one day.

VI.6.7. Public projects 1979.

Belmay Primary School. A steel frame covering the gable end of the school building, facing the main entrance and the street front, the public see each day, is so fashioned that the sculp-crete panels produced by the school can, each year, present a different theme, in line with satisfying the dynamic needs of education.

The first theme, done in conjunction with the 150th anniversary of the State of Western Australia, is composed of 72 smaller and some larger panels celebrating the unequalled variety of wildflowers found in the State. Each panel is done by one or two children giving the widest school participation.

The project was carried out entirely by the school staff and students, with a sculp-crete school kit, under the inspired leadership of Mr. Clive Elliot the enthusiastic principal.

Armadale Town Council Project. A 15 cm (6 inch) deep base relief eleven meters long depicting aspects of each ward of the town as requested by the Councillors representing the ward separated by tree trunks and unified with the frame of flowering gums. The life below ground and below the water line is depicted as well as that above.

VI.6.8. Free-standing articipation sculpture.

Commissioned by the Belmay Primary School to celebrate the Year of the Child. Criteria were to figure a teacher and group of children with which the live students could mingle, play and have feelings for and physical contact.

The success of the Sculpture is daily demonstrated to the delight of the principal, staff, students, parents and passers-by.

It's tough, non-perfectionist style (simulating real life) allows for repainting by the children, repairs and play galore.

A sense of belonging and individual attraction has been added to the school.

Goomalling District High School (primary and lower high school) in a small country town in Western Australia, the Sculpture commissioned by the Director General of Education as part of an architectural and community participation concept.

The Theme is Sheep and Wheat the sources of life in the town, complemented by the all important element of water, often absent during droughts. The theme was chosen through a community effort. The results indicate the sculp-crete methods versatility in producing the texture of wool, the pattern of wheat stubble, and a burned tree stump typical of Australian country areas. The Sculpture has as a back-drop the eight bas-relief scenes heat-carved by the students of the school, on moulds used for the casting of the sandwhich wall panels, of the Library, in that position.

VII INDUSTRALIZED BIO-FUNCTIONAL ECO-ARCHITECTURE

VII.1. Mass-production sandwich. Panels. (Floors, Walls).

Pragmatic experimentation showed that vertically cast elements, with concrete each side of a polystyrene core of about 10 cm, and each side of concrete taking its pattern and colour from a mould liner, within the steel formwork, acquire structural characteristic of considerable advantage as well as external and internal finishes. The vibration links the concrete and the polystyrene core with a 'root system', from each side, giving unusual resilience to impact blows onto the concrete. Further, laid horizontally, each thin concrete membrane acts as a flange and the polystyrene core as a web, structurally, fused through the "root system"

The thinness of the concrete facing membranes and the lightness of the core give a very light, highly insulating, strong, easy to handle panel.

Both production and building have now been carried out for a Government project in Western Australia which indicate the feasibility of a wide range of use for such panel systems. The specific use is described in Section VII.4.

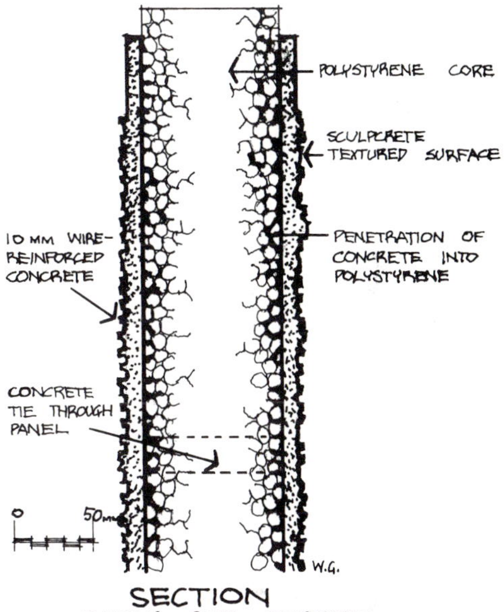

SECTION
SHOWING PANEL COMPOSITION

VII.2. City of Saudi Arabia.

An international firm of architects and kindred disciplines having designed buildings, and environment for a new city in Saudi Arabia were looking for an appropriate method to realize their intentions to give the traditional moslem art forms an appropriate expression to the city. An extended examination, of sculp-crete samples and the system, led them to recommend the sculp-crete method for the ceilings and other surfaces. The method recommended was mass production from mounted multiple elements, giving an individual character to each panel, through the imperfectionist nature of the sculp-crete technique itself.

VII.3. A Medical Consultants Residence.

The criteria for the house included a merging effect into the natural bush hill-side without undue fire danger; Exceptional insulation and use of sun shine in positive ways. (Perth is the sunniest capital city in the world, and the windiest.) The character of individual rooms to be established in the integral design as customary in some historic styles.

The design uses the concept of the bio-wall externally. This gives added insulation and the desired visual merging effect, as well as opportunity for planting for food and pleasure. The building panels have individual designs on the inside and the bio-wall troughs are cast integrally for the outside.

The choice of designs for individual rooms was developed in close conjunction with the clients. The price given for the construction was well below the normal cost of 'brick and tile' houses in the area. Construction 1980.

VII.4. Library Resources Centre Addition to District High School.

VII.1. Origin & Organisation.

Paul Ritter appointed as Architect by Public Works Department. Through the P.E.E.R. Institute and the co-operation of the Director General of the Department of Education who had reported, after a European study tour, on the desirability of greater community involvement in country schools in Australia.

Through a Government allowance of 5% to make the project into a community project through involvement, a major sculpture was obtained. Many agencies participated: State Education Department, Public Works Department, Schools Commission of the Federal Government, Local Member of Parliament, Shire Council, Services Clubs, Parents Citizen Association, School Secretary, Teachers, Students, 150th Celebration State Executive Committee, Local Commercial Firms, Health Department, Cleaners and Gardeners, Concrete and Cement Manufacturers, etc. Illustrated talks were given to the teachers, the Parents Citizen Assoc., the Goomalling Council, the citizens of Goomalling, the assembled school. Radio stations announced the project.

Through these agencies the following aspects of the project were carried out: over and above the building of the accommodation below average price:

a) teacher training in sculp-crete;
b) student design of sculp-crete decorative panels;
c) students creation of moulds for concrete casting;
d) competition for sculpture idea;
e) casting of students bas-relief panels and sculpture with fountain & landscaping;
f) educreational learning project by students producing a book for their library on the building processes involved;
g) the architect promised to return for a feedback session with the students end of year ceremony, a year later.
h) plaque commemorating Year Of The Child and 150th Anniversary of founding of Western Australia.

VII.4.2. Criteria for Design.

The building, in a high risk earthquake zone, had to be earthquake resistant. It had to fit into a pale yellowy-pink and orange red tile, existing school, forming one side of a courtyard.

A completely hinged construction, which allows every member to move without being damaged, solved the first problem. A white and grey, refined pavillion architecture, formed the complete contrast to complement the existing buildings giving the whole composition a focal point. The elements of wall and windows were kept to a small scale to harmonize with the architecture and the children.

VII.4.3. Roof — prefabricated

The roof made from universal mould form designed by P.E.E.R. Institute. This has special characteristics as a series of converging, sloping barrel roof sections. It is not only a very strong form but the mould can be simply adjusted so that sections to form any regular shape, from triangle to many sided or circle can be created from the same mould.

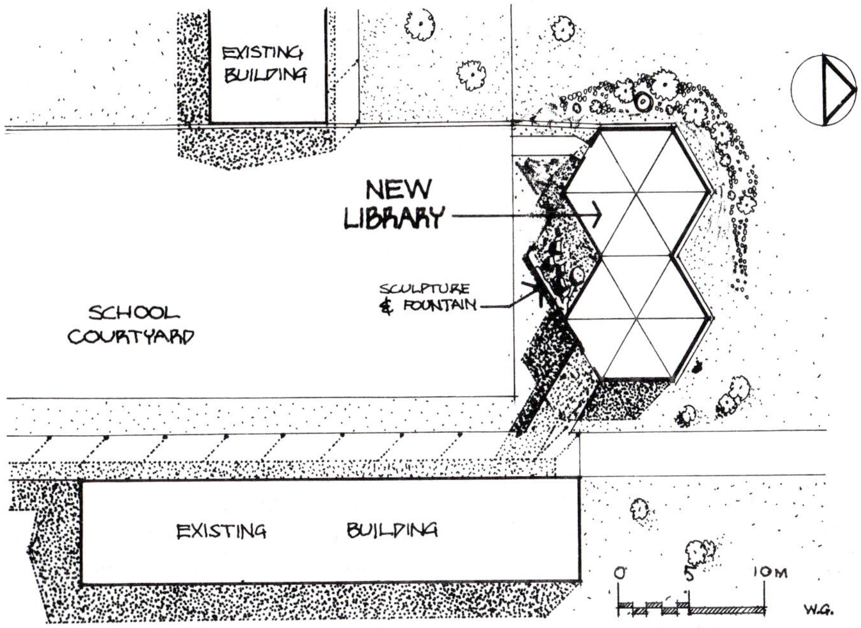

The strength of the shape had been tested long ago through a 1/2" thin asbestos roof, cast for testing purposes. This roof spanned 8 m without support and the Health Inspectors permitted its erection without the wire mesh that is obligatory under all normal asbestos covered roofs however small the span.

In this case the material for the roof was a sandwich with 5 cm insulation of polyurethane, with external and internal finish in fibreglass. Conduits were laid into the polyurethane during formation of the roof segments. The twelve triangular curved barrel vault units, making up the two, hexagons, covering the building, were brought to the site. They were set out, bolted together and hoisted onto the light steel columns, by crane, in less than an hour. A prototype had been tested in Government use for some seven years with perfect satisfaction.

VII.4.4. Walls and foundations forming plinth.

The walls were precast sandwich panels. A thin layer of concrete to give external and internal finish each side of a 10 cm polystyrene core. About 1 m wide, and 1.75 m high, their light weight allowed them to be handled easily by two men with a trolley.

The focal point of the school. Free standing sculpture by Robyn Cant, Simon Miller, Ralph Hibble integrated by architect Paul Ritter. Fountain made possible through local participation and donations.

Childrens sketches for bas-relief panels.

Joints were covered with bolted, fibreglass strips to give room for plenty of movement, to ensure waterproofing and to give neatness to panels, the edges of which are designed not to be perfect, but to allow easy handling without special care having to be taken. Where the students had provided the form liner the outside of the panels had the historic scenes of Goomalling which were the subject of the students memorial project. These were in colour and the water outlet of the fountain was incorporated to come out of a water-tank in the bas-relief.

This and all other ducting was easily accommodated at panel joint or in the polystyrene cores of the panels.

The sculp-crete pattern of random square spirals which gives the building a highly distinctive look was matched by the same pattern having been imprinted on the polystyrene formwork placed for pouring of the floor slab and edge beam. This is visible as a plinth on which the building stands.

The matching of precast and in situ concrete was extremely simple and gives a highly effective unity to the total structure.

VII.4.5. Micro-climate.

The hot dry climate with temperatures of 40 degree C. (over 100 degrees F.) and cool nights makes considerable demands on the designer. The Government regulations do not allow for air conditioning.

The very good insulation value of the polyurethane roof and the polystyrene sandwich walls are the basic safeguards. The windows are protected by fly wire and where sun might penetrate by mini-flyscreen sun-shade combination mesh. The windows facing in six directions can take advantage of cooling breezes from any direction. The high ceilings, suspended fans and extract vents at the apex of each hexagon ensure air-circulation of an unusually large amount of air for school rooms.

Natural lighting is all round and is excellent as is the reflected strip lighting round each hexagon.

The acoustic qualities of the ceiling are also highly satisfactory. Lack of time made it impossible to include the carpet heating which the P.E.E.R. Institute normally uses. This follows the logic of providing heating where it is most beneficial, low down, and from where it disperses most efficiently, warm air rising naturally. The heating provided is gas convector heating in each space. External grills protecting children against hot exhaust ducts are sculp-crete, matching the square spiral theme of walls and plinth.

Internal finish purposely random greys on pattern to obviate the "getting dirty" effects.

External finish to encourage micro-growth.

Watercourse pattern to create lively sound and light effects.

On site cast concrete plinth matches pre-cast panels.

Entrance from link leading to school veranda.

VIII EPILOGUE

If this has been an introduction to an architecture for the third millenium then ours are some of the earliest fumbling efforts. The reader should not be put off by their crudeness, insufficiencies or difference in taste. Do not throw the baby out with the bathwater, as we do so often when the first interpretation or realization of an idea or invention is produced. Prototypes, by their very nature, tend to have many weaknesses. Spot them and put them right. Do not condemn the idea or the concepts because of an interpretation. Sculp-crete and bio-functional architecture is not something you can see and judge at this point. The evidence suggest that its basis is valid and that its flexibility and open endedness allows development and interpretations to your own heart's content.